THE EAST-WEST CENTER—formally known as "The Center for Cultural and Technical Interchange Between East and West"—was established in Hawaii by the United States Congress in 1960. As a national educational institution in cooperation with the University of Hawaii, the Center has the mandated goal "to promote better relations and understanding between the United States and the nations of Asia and the Pacific through cooperative study, training, and research."

Each year about 2,000 men and women from the United States and some 40 countries and territories of Asia and the Pacific area work and study together with a multinational East-West Center staff in wide-ranging programs dealing with problems of mutual East-West concern. Participants are supported by federal scholarships and grants, supplemented in some fields by contributions from Asian/Pacific governments and private foundations.

Center programs are conducted by the East-West Communication Institute, the East-West Culture Learning Institute, the East-West Food Institute, the East-West Population Institute, and the East-West Technology and Development Institute. Open Grants are awarded to provide scope for educational and research innovation, including a program in humanities and the arts.

East-West Center Books are published by The University Press of Hawaii to further the Center's aims and programs.

Computations of Surface Layer Air Parcel Trajectories, and Weather, in the Oceanic Tropics

INTERNATIONAL INDIAN OCEAN EXPEDITION
METEOROLOGICAL MONOGRAPHS *Number 7*

The International Indian Ocean Expedition Meteorological Monographs, published as East-West Center Books by The University Press of Hawaii, contain detailed discussions and supporting data on the various components of the general atmospheric circulation over the Indian Ocean, as well as the results from measurements of atmosphere-ocean interaction made as part of the expedition's observational program.

COMPUTATIONS OF SURFACE LAYER AIR PARCEL TRAJECTORIES, AND WEATHER, IN THE OCEANIC TROPICS

Adrian H. Gordon
Ronald C. Taylor

AN EAST-WEST CENTER BOOK
The University Press of Hawaii
Honolulu

The publication of this volume has been aided by Grant No. GA-386 from the National Science Foundation.

Contribution No. 71–9 of the Department of Meteorology, University of Hawaii

Manufactured in the United States of America

Library of Congress Cataloging in Publication Data

Gordon, Adrian Hugo.
Computations of surface layer air parcel trajectories, and weather, in the oceanic tropics.

(International Indian Ocean Expedition meteorological monographs; no. 7) (Contribution of the Department of Meteorology, University of Hawaii; no. 71–9)
"An East-West Center book."
Includes bibliographical references.
1. Boundary layer (Meteorology) 2. Dynamic meteorology. 3. Tropics. 4. Indian Ocean.
I. Taylor, Ronald C., joint author. II. Title.
III. Series. IV. Series: Hawaii. University, Honolulu. Dept. of Meteorology. Contribution of the Department of Meteorology, University of Hawaii; no. 71–9.
QC924.I5I5 no. 7 [QC880.4.B65] 551.6′5′165s [551.5]
ISBN 0–8248–0253–5 72–92065

Contents

I am still uncertain what to do with the winds that appear to cross the equator. . . . I always find my pen sticks in the paper and refuses to move when I try to draw an isobar across the equator.

Sir Napier Shaw
The Air and its Ways, 1923

To put the point in a crude form, I do not know whether, in practice, the winds have to adjust themselves to the pressure conditions, or the pressure distribution is the result of the motion of the air.

Sir Napier Shaw
Introduction to Gold's *Barometric Gradient and Wind Force*, 1908

I. *Introduction*

Can we account for the following observation? In India, during the months of the summer monsoons, satellite observations show striking variations of cloudiness over comparatively short distances, with a tendency for the clouds to be arranged in nearly east-west bands: for example, massive cloudiness between the equator and 5 degrees north; clear between the equator and 5 south. Further, these patterns persist for many days in a steady flow of air over the mid-Indian Ocean. An examination of the surface synoptic chart for the same period shows the air drifting across the isobars toward the lower pressure north of the equator with, again, the isobars directed more-or-less east and west.

Equally striking features are observed in the low-level distribution of cloudiness extending over the tropical Pacific and in the distribution of rainfall over the atolls of the Gilbert-Ellice and Line islands. Though we may lack confidence, on a particular occasion, about or concerning relevant details of either the fields of pressure or motion, we often find rather "bland"-looking pressure fields apparently producing rather striking features in the field of motion.

Leaving aside for the moment the perennial question of data deficiencies, how are we to account for these features of the motion field in terms of the pressure field, from which the air derives its kinetic energy?

The question, of course, is not a new one; it was a central theme of meteorological inquiry nearly a century ago. One empirical solution offered by Abercromby (*1*) depicting the distribution of clouds and weather around a composite mid-latitude depression became the basis of successful forecasting in the United Kingdom around the turn of the century. This approach, of course, is successful in middle latitudes since the pressure field is a reasonably good first approximation to the field of motion.

The quality of this approximation degrades rather rapidly with decreasing latitude, though it is not impossible to expect it to hold on some occasions. The general validity of such reasoning, however, has yet to be demonstrated with conviction. Indeed, rather than persevere in this dynamical direction of pressure-wind relations, tropical investigators more and more focused their attention on just the field of motion, attempting to relate the distribution of clouds and weather to the directly computed horizontal divergence. This they did with some success, but only with confidence when the observational network was considerably expanded, as it was during selected periods, namely nuclear weapons testing.

With the modest expansion of observational networks in low latitudes in recent years, the habit of analyzing the field of motion, to the near exclusion of the pressure field, has been widely adopted. However, within the last few years tropical forecasters in Asia and Africa have expressed growing doubts over the usefulness of just the field of motion as a forecasting tool and have again directed their attention to the pressure field.

This is the approach we adopt here in an effort to understand the questions outlined in the first few paragraphs. Starting with a specified (steady) pressure field, a frictional constraint, and, implicitly, a variable Coriolis acceleration, we offer a first attempt to compute the field of motion in the near surface layer and its divergence. We compute the paths of individual parcels, treating them as steady streamlines under suitable assumptions, and examine the results diagnostically.

It appears that many of the puzzling features of cloud and weather distributions can be accounted for by the divergence of the pressure field and its implications in how moving parcels of air change their horizontal divergence as they move about.

II. *Historical Background*

One of the lasting achievements of mathematicians and physicists during the last half of the eighteenth century was the introduction of the theory of the motion of a continuum and two basically different ways of describing it dynamically. Previously, several useful results—fluid pressure, for example—had been obtained by considering fluid media to be a collection of Newtonian particles undergoing elastic collisions between themselves and their surroundings. Such a conceptual scheme, in the hands of Jean and Daniel Bernoulli, for example, failed to give additional results of general interest, and not until the work of Euler, d'Alembert, and Lagrange were results of wide application available in the form of the general field equations of hydrodynamics, the equation of continuity, and the so-called Lagrangian equations of fluid media.

The description of continuous media in terms of field equations has sustained an enormous impact upon physical theory over the past two hundred years. This technique of specifying, say, the continuum velocity at fixed locations throughout a region of space at a specified time is almost the sole contribution of Leonhard Euler (*2*), undoubtedly the greatest analyst of his generation. Earlier, however, d'Alembert made use of the field description for a particular problem, but did not generalize his technique. Further, Euler provided an alternative description to that of his field equations in a form of what mistakenly is generally referred to as the Lagrangian description! Euler did not perfect this technique of describing some dynamical feature of one particle and following its history for subsequent time, but he did originate the method, including the elegant formula describing parcel mass continuity,

$$\operatorname{div} v = \frac{1}{J}\frac{dJ}{dt}$$

where J is the Jacobian expressing the point transformation between initial and subsequent coordinates, though this refinement was the contribution of Carl G. J. Jacobi some fifty years later. The particle interpretation was, of course, very much in the spirit of Newtonian mechanics, though little was gained mathematically since both descriptions, the field and parcel techniques, are essentially nonlinear. The parcel technique, through a longstanding neglect of Euler's work, is generally associated with the name of Lagrange, who perfected it in his famous book, *Mechanique Analytique* (*3*); the book itself contains no reference to the earlier work of Euler. In fact, this error was not corrected during Lagrange's lifetime, and only in the third edition (1853) edited by Joseph Bertrand is a footnote finally added to correct the historical record. By this time the precedent had been established and subsequent investigations entailing the dynamics of fluid parcels were henceforth labeled as Lagrange's technique. Clearly, it would be more appropriate to identify the parcel method as the Euler-Lagrange technique. Nevertheless, with few exceptions—namely, Sommerfeld (*4*)— Euler's early development of the parcel equations is almost universally identified with the name of Lagrange.

One of the first applications of the "Lagrangian technique," which held considerable physical interest, was offered by Franz Gerstner (*5*) who investigated steady-state water waves. Gerstner's solution was criticized because his waves were rotational, though he was one of the first to show that his solutions correspond to a possible fluid motion since they satisfied the parcel equation of continuity originally derived by Euler. A modern discussion of Gerstner's troncoidal waves is provided by Milne-Thomson (*6*).

The application of the Lagrangian technique to the study of atmospheric motions is quite recent; nevertheless, it is useful to recall that the earliest synoptic investigations by Brandes (*7*), Espy (*8*), and others—all using a field description—generally expressed their results in terms of air parcels flowing along particular "trajectories." Indeed, the distinction between a streamline and a trajectory in meteorological discussions was, at best, obscure; it remained so until after the turn of the century.

The first, and still memorable, application of the Lagrangian technique was the extensive investigation of surface-air trajectories by Shaw and Lempfert (*9*), but before considering their results it is fitting to consider a simpler dynamical problem: the motion of a free particle on a rotating sphere.

Coriolis (*10*), through his interest in waterwheels (!), contributed three fundamental papers of wide interest concerning rotating coordinate systems, but he never considered any geophysical applications. A year later, Poisson (*11*), being interested in projectile motion, contributed two papers that included detailed discussions of the motion of a free particle relative to a rotating sphere, though again no particular geophysical interpretations were offered.

In the light of the results published between 1803 and 1838 by Coriolis, Poisson, and others, it is indeed surprising to find three investigations by Von Baeyer (*12*), Ohlert (*13*), and Mousson (*14*), who computed inertial trajectories of air parcels through an appeal to the discredited Hadleian idea of angular momentum conservation! Though Von Baeyer expressed some skepticism about Hadley's principle, his "dynamical" courage finally deserted him and he ultimately returned to Hadley's formulation; both Ohlert and Mousson, on the other hand, did not express any doubts concerning the basic validity of this principle. These dynamical discussions were finally corrected by Sprung (*15*) who appealed to the results of Coriolis, but confined his discussion to inertial motion near the pole where the normal elliptic integrals are greatly simplified. Regarding the wider problem, where the elliptic integrals must be evaluated, Sprung noted that ". . . the solution of this problem . . . does not seem to be worth the while . . . because . . . the careful determination of the path has only a subordinate interest in meteorology, since the motion . . . that the particles of air actually follow, the 'inertia path,' has been completely refuted by the synoptic weather charts" (p. 143). Apparently, Sprung did not think that a knowledge of the actual inertial paths would lead to a deeper understanding of the forces producing motions. This attitude, however, was later criticized by Whipple, but before discussing his results, it is worth remembering a nearly forgotten book on atmospheric dynamics published by Cordeiro (*16*). In his chapter on "Motion Relative to the Earth" the author presented several results of inertial motion derived from a simple physical understanding of salient forces, along with a reasonably direct mathematical description. He obtained correct results for inertial trajectories in both high and low latitudes without resorting to the evaluation of elliptic integrals. Cordeiro did not include references to previous works, and he does not appear to have published any other meteorological investigations; his book was reviewed by Gold (*17*).

In contrast to Cordeiro's approach, Whipple (*18*) provided a reasonably complete mathematical discussion of inertial motion on a smoothly rotating sphere, and took Sprung to task for not completing the problem the latter set out to solve. Whipple included seven distinct cases of inertial trajectories (based upon values of various arbitrary constants) and noted that "the study of motion under no forces should precede the theoretical study of motion governed by pressure" (p. 457). Here, there is a distinct resounding of the views of Napier Shaw, who, incidentally, submitted Whipple's paper to the *Philosophical Magazine*.

As noted above, the distinction between a streamline and a trajectory was not widely appreciated in meteorological discussions of air motions, say, seventy years ago. Though such expressions as "paths of air" and "the path of air motions" were crudely related to trajectories, it is clear from the context of many discussions that the latter were often confused with streamlines. It is therefore not surprising that Napier Shaw (*19*) thought sufficiently of this state of affairs to make the distinction explicit—about the time he and Lempfert were beginning their investigation of *The Life History of Surface Air Currents*—and to demonstrate some of its mathematical consequences. Moreover, Shaw encouraged Hudson (*20*) to investigate additional mathematical consequences of the combined effect of a moving circular depression and air motions relative to its center; the resulting trajectories Hudson called "anemoids."

The rather prosaic frontispiece in Shaw and Lempfert's (*9*) memorable paper immediately indicates that something new is being offered in the matter of examining the motion of the atmosphere. Here we see eight graphs, depicting various classes of trajectories, that display changes of temperature, pressure, and speed as a function of time—in short, the "life history of surface air currents." Shaw and Lempfert constructed their trajectories from hourly observations of wind and pressure made at stations throughout the British Isles and the adjacent regions of western Europe; several trajectories, covering the width of the Atlantic, were constructed from 1882 synoptic charts.

The major dynamical findings of this investigation were expressed by Shaw in his introduction to Gold's (*21*) memoir two years later: ". . . one of the most obvious conclusions from that research in the kinematics of the atmosphere, is that, as a matter of fact, air in its motion is subject to very little and very slow change of velocity. The meteorological time unit is the hour or the day, and in these intervals the changes in the velocity of the moving air are comparatively small . . ." (pp. 6–7). Five years later Shaw (*22*) provided a formal discussion of this conclusion in the form of his "strophic balance," and coined a new term for the first edition of the *Meteorological Glossary* (*23*), the "geostrophic" wind.

The familiar geostrophic formula appears in the literature nearly forty years earlier, though its full dynamical significance was hardly appreciated; rather, it appears, in the hands of Mohn (*24*) and Peslin (*25*), as little more than an empirical formula of uncertain value. Further, these latter investigations were published in rather obscure places, and neither appears to have attracted much notice; indeed, Shaw and Lempfert do not refer to either paper, nor does Sprung (*26*) in his otherwise thorough book. Finally, in the Introduction to Gold's (*21*) memoir Napier Shaw suggested the next step and asserted that subsequent investigations of wind and pressure ought to start from the "strophic balance," and that departures from such a balance "are likely to be productive as a subject of meteorological study" (p. 11).

Many synoptic investigators prior to the work of Shaw and Lempfert attempted to relate the observed weather to particular features of the pressure field, that is, properties of the field of motion were related to the force field. The obscure physics of this procedure must have impressed Shaw considerably, especially in the light of his admiration for Clerk Maxwell's work, but, in any event, he and Lempfert supplied a vital ingredient to a possible solution by relating the weather to a consequence of the field of motion—the "horizontal dilation of areas"—on the Lagrangian plan. Shaw and Lempfert, with the aid of G. T. Bennet's technique of computing divergence from the wind field, were able to present models of the distribution of clouds and precipitation with respect to the center of a typical depression; and, what is more, were able to show why such a distribution ought to be expected.

The provocative results of Shaw and Lempfert appear to have stimulated a wide interest in particle dynamics among continental meteorologists as well as those in Britain. Two years after their monograph was published Gold (*27*) determined particle trajectories in both zonal and circular isobaric fields, deriving, respectively, the well-known cycloid and cardioid curves. In addition, with Napier Shaw's urging, Gold attempted to compute the time necessary for the geostrophic balance to be achieved by a parcel starting from rest in a pressure field of zonal isobars; in his words ". . . whether the pressure is likely, in actual circumstances, to continue steady long enough for such a state of affairs to come about (i.e., geostrophic balance). We can get some idea of the time required to lapse by considering the motion of a particle on the earth's surface (starting from rest), moving under a constant force in a constant direction" (p. 13). Gold considered the geostrophic wind to exist when the parcel was moving parallel to the isobars, and computed this time to be about 4 hours at 50°N. Unfortunately, though the parcel does indeed move parallel to the isobars in about this time, it has exactly twice the geostrophic speed; in fact, for time-independent pressure fields, the parcel never attains the geostrophic speed unless it starts with it. This is a curious error in a memoir that is otherwise thorough and clearly not lacking in the relevant mathematical framework. Nevertheless, Gold's memoir is a milestone in the development of dynamic meteorology, reporting the first comparison between observed winds at 1,000 meters and the geostrophic wind deduced from the surface pressure, and discussing for the first time the physical significance implied by the quadratic nature of the gradient wind equation, and partial hinting at what later became known as the thermal wind equation. Finally, Napier Shaw's introduction to Gold's memoir is a succinct statement of the whole problem of wind and pressure, hardly outdated sixty years later.

Continental meteorologists contributed to the growing narrative on particle dynamics beginning with Sandstrom's (*28*) instructive discussion of balanced flow with (surface) friction, following the original formulation of Guldberg and Mohn (*29*). Sandstrom considered some possible frictional trajectories, using data from (Scandinavian) synoptic charts, and developed a plausible physical argument to show that the expected parcel air motions should be "oscillatory paths"; he then selected some synoptic (winter) situations in which the pressure field was sensibly steady and depicted the actual field of motion using his recently developed isogonisotach technique. Appealing to the near steady character of his chosen synoptic charts, he identified the resulting streamline map with that of actual trajectories, and thereby demonstrated the reasonableness of his initial deduction, namely, that the trajectories do indeed follow "oscillatory paths." Whether or not the Guldberg and Mohn formulation of friction, including Sandstrom's own ingenious modifications, is entirely adequate to the particular geography at hand, Sandstrom's formulation undoubtedly contains the salient physics of the situation. Finally, Sandstrom's paper contains no reference to the works of Shaw and Lempfert (*9*), or Gold (*21*).

Felix Exner (*30*) clearly found Sandstrom's results of sufficient interest to warrant a careful theoretical discussion based directly upon the (horizontal) equations of motion. In particular, he presented the two-dimensional particle solution for the case of east-west isobars, constant Coriolis force, and time-independent meridional pressure field, showing how parcel trajectories (without friction) near 50°N crudely resembled those obtained by Sandstrom. Exner also presented the particle solution for the same case, with friction, following the Guldberg and Mohn formulation; however, he confined his remarks largely to the inadequacies of the frictional formulation. Finally, Exner (*31*) provided an extended discussion of particle dynamics in his well-known book, though he did not repeat it in the second edition (*32*).

The lack of any considerable discussion on the part of Exner (*30*) of the parcel equations with friction was soon remedied by Hesselberg (*33*), who also included some discussion of time-dependent pressure fields. Unlike Exner he provided no diagrams of computed trajectories, but only a qualitative discussion. Nevertheless, his results did include the complete particle solutions and he clearly acknowledged the importance of initial conditions. If friction is neglected and the pressure force depends only on time, then his results reduce to a complete particle dynamics of nonviscous flow in a horizontal plane. Thirty-five years later Forsythe (*34*), unaware of Hesselberg's paper, followed the same approach in his discussion of time-dependent pressure fields. Lastly, Hesselberg acknowledged the previous work of Gold (*21*) and Sandstrom (*28*), as well as Exner (*30*).

The next twenty-five years (1915–1940) witnessed the first attempts to examine the atmosphere in direct Lagrangian terms, beginning with Meisinger's (*35*) manned "constant elevation free balloon flights from Fort Omaha." In these experiments, in which he eventually lost his life, Meisinger attempted to compare the observed balloon trajectory with features of the surface pressure distribution. On a much smaller scale, Richardson and Proctor (*36*) used some Lagrangian results of the dispersal of toy balloons while investigating features of atmospheric diffusion in the surface layer. Finally, Ackerman and Piccard (*37*) reported the first (unmanned) constant-level-balloon observations made in the United States using balloons constructed of cellophane. Subsequent developments along this line are summarized in Angell's (*38*) extensive review.

The increased attention given to new observational techniques for obtaining Lagrangian data and to the interpretation of those data, as well as the equally significant results of Eulerian perturbation analysis, motivated by the polar front theory, appears to have encouraged Bjerknes' (*39*) theoretical discussion of both techniques. He discussed the advantages to be gained from a Lagrangian analysis of some simple fluid models, though his paper appears to have gained little notice even among theoreticians.

Just prior to World War II German meteorologists took up several aspects of particle dynamics, and, apparently, rediscovered the early work of Shaw and Lempfert. Significant contributions were published by Baur and Philipps (*40*), Philipps (*41*), and an interesting, though largely ignored, investigation by Lotz (*42*); further Koschmieder (*43*), in the first edition (1933) of his well-known book, provided a useful summary of earlier results though he deleted, as Exner did, most of the material in the second edition (1941).

Baur and Philipps (*40*), though not appealing to the complete solutions of the parcel equations, formulated the problem in terms of balanced flow with friction; they provided numerous graphs giving the pressure-gradient force as a function of the angle between isobars and the wind vector for various values of wind speed over the sea. Following Sandstrom's (*28*) suggestion for modifying Guldberg and Mohn's formulation of friction, but more importantly following Ekman's (*44*) results, Baur and Philipps' resultant friction vector is directed 45 degrees from the direction of motion rather than just opposing it, as in the original formulation of Guldberg and Mohn. Two years later Lotz (*42*), in the spirit of Shaw and Lempfert, determined the horizontal velocity divergence from synoptic surface charts by following three sample parcels defining a triangular area. Lotz made a clear distinction between geostrophic trajectories and his "trajekorien des Windes erster Näherung," choosing his first approximation by attempting to evaluate the geostrophic departure, a technique later utilized (independently) by Franceschini and Freeman (*45*), and Franceschini (*46*).

In the early 1950s constant-pressure balloons were perfected and flown near the 300 mb level over North America and the Pacific Ocean, providing the first substantial sample of Lagrangian meteorological data on a large scale (*38*). The success of this new tool stimulated renewed interest in particle dynamics, including an attempt to evaluate the (mean) Lagrangian frictional force at 300 mb, and the nonlinear oscillations of a particle in a long wave (*47*). Finally, Wurtele (*48*) demonstrated some interesting theoretical advantages to be gained from a Lagrangian approach to some familiar dynamical problems.

Suspecting that certain defects in numerical forecasts resulted from the Eulerian technique itself, Wiin-Nielsen (*49*) adopted a trajectory approach and applied it to both barotropic and baroclinic models. Other Lagrangian approaches, adapted to numerical prediction, have been advanced by Djuric (*50*) and Okland (*51*).

III. *Theoretical Development*

A. FRICTIONLESS FLOW

The equations of motion for horizontal frictionless flow referred to a rotating tangent plane coordinate system are

$$\begin{aligned} \frac{du}{dt} - fv &= -\frac{1}{\rho}\frac{\partial P}{\partial x} = P_x \\ \frac{dv}{dt} + fu &= -\frac{1}{\rho}\frac{\partial P}{\partial y} = P_y \end{aligned} \tag{1}$$

where the above and ensuing symbols are as given in the List of Mathematical Symbols. Gordon and Shaw (*52*) demonstrated that within the tropics the additional terms included in a spherical or curvilinear coordinate system may be neglected as they are at least one order of magnitude less than those contained in Equation (1) (*52*).

Equation (1) possesses unique solutions for constant values of f, P_x and P_y. Thus, (1) may be written

$$\begin{aligned} \dot{u} - fv &= P_x \\ \dot{v} + fu &= P_y. \end{aligned} \tag{2}$$

Writing the first relation in terms of v

$$v = \frac{\dot{u} - P_x}{f} \text{ and differentiating, } \dot{v} = \frac{\ddot{u}}{f}.$$

Substitution of $\dot{v}$ in the second relation of (2) gives

$$\ddot{u} + f(fu - P_y) = 0.$$

If we let $fu - P_y = m$ the relation above becomes

$$\ddot{m} + f^2 m = 0. \tag{3}$$

Equation (3) is the equation of simple harmonic motion and possesses a solution of type

$$m = c_1 \cos ft + c_2 \sin ft. \tag{4}$$

Substituting for $m = fu - P_y$ in (4) and using the relation

$$v = \frac{\dot{u} - P_x}{f} \qquad \text{we have}$$

$$u = \frac{m + P_y}{f} = c_1 \cos ft + c_2 \sin ft + P_y \tag{5}$$

$$v = \frac{1}{f}(-c_1 \sin ft + c_2 \cos ft - P_x).$$

To evaluate the constants of integration let $u = u_0$, $v = v_0$ at $t = 0$; that is, u_0, v_0 are the components of the wind velocity at the initial time of observation. Then,

$$u_0 = \frac{c_1 + P_y}{f} \qquad \text{and} \qquad c_1 = fu_0 - P_y$$

also,

$$v_0 = \frac{c_2 - P_x}{f} \qquad \text{and} \qquad c_2 = fv_0 + P_x.$$

Now substituting for c_1 and c_2 in (5) we obtain

$$\begin{aligned} u &= \left(u_0 - \frac{P_y}{f}\right)\cos ft + \left(v_0 + \frac{P_x}{f}\right)\sin ft + \frac{P_y}{f} \\ v &= \left(v_0 + \frac{P_x}{f}\right)\cos ft - \left(u_0 - \frac{P_y}{f}\right)\sin ft - \frac{P_x}{f}. \end{aligned} \tag{6}$$

Now integrating (6) it follows that

$$\begin{aligned} x &= \frac{1}{f}\left[P_y t + \left(u_0 - \frac{P_y}{f}\right)\sin ft + \left(v_0 + \frac{P_x}{f}\right)(1 - \cos ft)\right] + x_0 \\ y &= \frac{1}{f}\left[-P_x t + \left(v_0 + \frac{P_x}{f}\right)\sin ft - \left(u_0 - \frac{P_y}{f}\right)(1 - \cos ft)\right] + y_0 \end{aligned} \tag{7}$$

where x_0, y_0 are the initial coordinates of the parcel at the beginning of the integration period.

Before considering trajectories of air parcels computed from the numerical integration of (6) and (7) for selected cases of synoptic and climatological scales of motion, it will be of interest to discuss briefly some special simplified solutions of these equations.

Firstly, we may write (6) in the form

$$\begin{aligned} u &= u_0 \cos ft + v_0 \sin ft \\ v &= v_0 \cos ft - u_0 \sin ft. \end{aligned} \tag{8}$$

Squaring and adding the two relations in (8) we have the diagnostic equation

$$(u - u_g)^2 + (v - v_g)^2 = (u_0 - u_g)^2 + (v_0 - v_g)^2. \tag{9}$$

One solution is

$$\begin{aligned} u - u_g &= u_0 - u_g \\ v - v_g &= v_0 - v_g \end{aligned}$$

or

$$\begin{aligned} u &= u_0 - u_g \\ v &= v_0 - v_g. \end{aligned}$$

Thus one solution of the diagnostic equation (9) is that the wind is not accelerated. This can only occur when the wind is geostrophic. No other specific solution can be found from the other root. When any other initial wind occurs the solution is given by the time-dependent equations (6).

1. Inertial Flow

We may let P_x and P_y both equal zero in (6) and (7) retaining the initial velocity components u_0, v_0. The flow is then inertial and in the absence of dissipating forces will obviously continue with constant speed V_0 where $V_0 = (u_0^2 + v_0^2)^{1/2}$. Thus (6) reduces to

$$\begin{aligned} u &= u_0 \cos ft + v_0 \sin ft \\ v &= v_0 \cos ft - u_0 \sin ft \end{aligned} \tag{10}$$

from which the relation $V_0 = (u_0^2 + v_0^2)^{1/2}$ follows. Equation (7) reduces to

$$x = \frac{u_0}{f} \sin ft + \frac{v_0}{f}(1 - \cos ft)$$
$$y = \frac{v_0}{f} \sin ft - \frac{u_0}{f}(1 - \cos ft) \tag{11}$$

when the motion commences from the origin of the coordinate system. If f is held constant the trajectories given by (11) are inertial circles of radius V_0/f. This result may be obtained quite simply by equating the Coriolis and centrifugal accelerations to one another.

$V_0^2/R = f V_0$, or $R = V_0/f$ where R is the radius of the circle of inertia. If f is allowed to vary in accordance with the function $2\Omega \sin \phi$ various inertial curves result. These will be referred to presently.

2. Meridional Pressure Gradient and Coriolis Parameter, Both Invariant with Latitude

We may align the isobars with the x axis of our coordinate system. Then $P_x = 0$ and (7) becomes

$$x = \frac{1}{f}\left[P_y t + \left(u_0 - \frac{P_y}{f}\right)\sin ft + v_0(1 - \cos ft)\right] + x_0$$
$$y = \frac{1}{f}\left[v_0 \sin ft - \left(u_0 - \frac{P_y}{f}\right)(1 - \cos ft)\right] + y_0. \tag{12}$$

A special case of the trajectories for a constant meridional pressure gradient occurs for the initial conditions $u_0 = 0$, $v_0 = 0$, $x_0 = 0$, $y_0 = 0$. Then (12) reduces to

$$x = \frac{P_y}{f}\left(t - \frac{1}{f}\sin ft\right)$$
$$y = \frac{P_y}{f}(1 - \cos ft). \tag{13}$$

Equation (13) is the equation of a cycloid.

3. Other Cases Where the Meridional Pressure Gradient and Coriolis Parameter Are Variable

Equations (6) and (7) and the special cases of these equations for inertial flow and for a pressure gradient directed along the y axis that have been discussed so far have been derived for a constant Coriolis parameter. Although a reasonable approximation may be made for air flow in temperate latitudes by assuming that $\partial f/\partial y$ is negligible, this cannot be done in low latitudes where f is much smaller and $\partial f/\partial y$ is larger than in middle and high latitudes. The solutions of the equations of motion in a spherical coordinate system include the variation of the Coriolis parameter, since it is a function of ϕ, the latitude coordinate of the system. In a Cartesian coordinate system a rough approximation may be made for latitudes in the tropics by using the relation $f = 2\Omega y/a$, where a is the radius of the earth, and y is measured from the equator.

Substituting the above term in (1) for the special case when $P_x = 0$ we have, from the first relation

$$\ddot{x} = \frac{2\Omega y\dot{y}}{a},$$

then

$$u = \frac{\Omega}{a}(y^2 - y_0^2) + u_0 \tag{14}$$

and, substituting this in the second relation of (1), we have

$$\ddot{y} + \frac{2\Omega y}{a}\left[\frac{\Omega}{a}(y^2 - y_0^2) + u_0\right] = P_y. \tag{15}$$

Multiplying through by $\dot{y}$ enables (14) to be readily integrated. If we insert the boundary condition $v = v_0$ at $t = 0$

$$v^2 = v_0^2 - \frac{\Omega^2}{a^2}(y^2 - y_0^2)^2 - \frac{2\Omega u_0}{a}(y^2 - y_0^2) + P_y\,(y - y_0). \tag{16}$$

Component velocities could be calculated from (14) and (16) for specified intervals of the y coordinate. Since $dy/dx = v/u$

$$x_n = x_{(n-1)} + \frac{u_{n-1} + u_n}{v_{(n-1)} + v_n}(\Delta y) \tag{17}$$

x_n is the x coordinate corresponding to the specified y coordinate for which u_n and v_n are calculated from (14) and (16). Trajectories of the motion may then be calculated from (14), (16), and (17), where Δy is some specified interval of distance in y direction.

A special case of the latter equations occurs when air starts its trajectory at the equator from rest. Then

$$u = \frac{\Omega y^2}{a}$$

$$v = \left[2P_y\,y - \frac{\Omega^2 y^4}{a^2}\right]^{1/2}.$$

The peak of the amplitude of the trajectory occurs when $v = 0$. At this point of the curve

$$y = \left(\frac{2P_y\,a^2}{\Omega^2}\right)^{1/3}. \tag{18}$$

Figure 1 shows the latitudes given by (18) for air leaving the equator from rest and traveling poleward under different constant meridional pressure-gradient accelerations. A value of P_y of 0.016 cm sec^{-2} is approximately equal to 1 millibar per 5 degrees of latitude at the surface. It can be shown that a gradient as small as 0.0625 mb/5° of latitude will produce a latitude displacement of nearly 5 degrees. On the other hand a pressure gradient 80 times as great (5 mb/5° of latitude) will cause a northward displacement of only 21 degrees of latitude.

Equation (16) may also be solved for air parcels starting from rest from other latitudes. It then takes the form

$$v = \left[2P_y\,(y - y_0) - \frac{\Omega^2}{a^2}(y^2 - y_0^2)^2\right]^{1/2}. \tag{19}$$

The peaks of the trajectories will then be given by setting $v = 0$ and solving for y for different pressure gradients and initial starting latitudes

$$(y - y_0)\,(y + y_0)^2 = \frac{-2P_y\,a^2}{\Omega^2}. \tag{20}$$

Figure 2 shows the peak latitudes reached by air parcels starting from rest from different latitudes between the equator and 20° latitude with a poleward directed meridional pressure gradient of 0.016 cm/sec^2 (1 mb per 5° of latitude as measured at the equator). It is seen that there

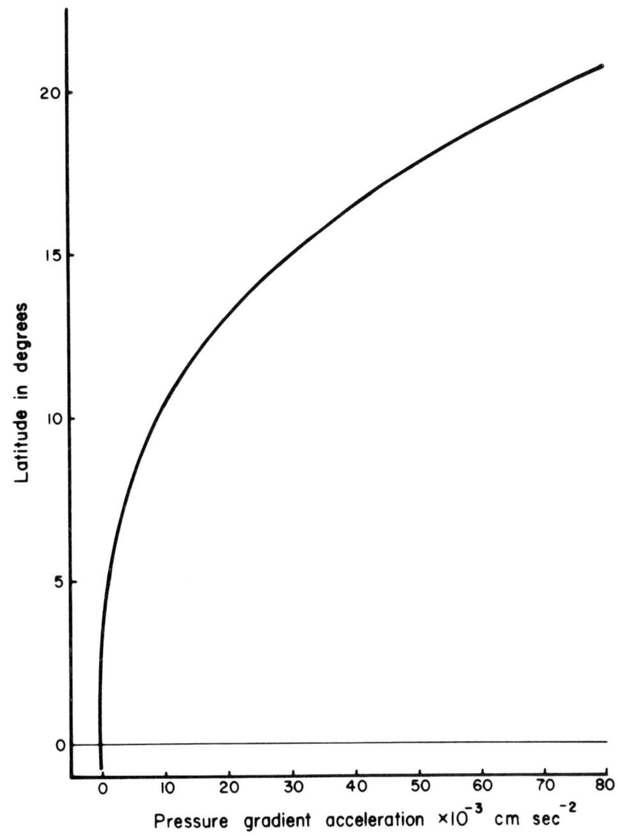

FIGURE 1. Maximum latitudes reached by parcels leaving the equator from rest as a function of P_y, the meridional pressure gradient.

is considerable convergence of the air parcels at the peaks of the trajectories. This reaches a maximum for air parcels leaving from 4 degrees of latitude for the pressure gradient chosen. The calculation of the divergence from the trajectories will be discussed in a later section.

The integrations of the equations of motion carried out in the previous sections have all been performed with a constant pressure gradient. An interesting case may be examined in which the meridional pressure gradient is proportional to the y coordinate. We maintain the Coriolis parameter constant.

Let us write the equations (1) in the form

$$\begin{aligned} \frac{du}{dt} - fv &= 0 \\ \frac{dv}{dt} + fu &= \alpha^2 y \end{aligned} \tag{21}$$

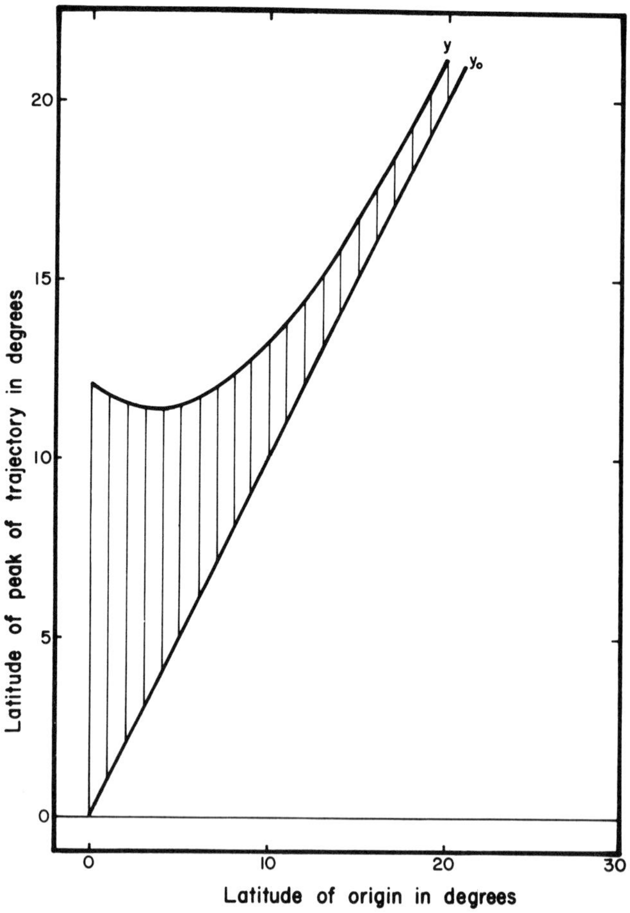

FIGURE 2. Maximum latitudes reached by parcels starting from rest from different latitudes with a constant meridional pressure gradient of 1 millibar per 5 degrees of latitude (P_y = 0.016 cm sec^{-2}).

where α^2 is a proportionality constant with dimensions sec^{-2}. We have $\ddot{v} + n^2 v = 0$, where $n^2 = f^2 - \alpha^2$, the solution of which is, for $\alpha < f$,

$$v = v_0 \cos nt + \frac{f}{n}\left(\frac{\alpha^2 y_0}{f} - u_0\right)\sin nt \tag{22}$$

where u_0, v_0 are the initial velocities and y_0 is the initial y coordinate. Also,

$$u = \frac{1}{f}\left[\alpha^2 y + v_0 \sin nt - f\left(\frac{\alpha^2 y_0}{f} - u_0\right)\cos nt\right]. \tag{23}$$

Integration of (22) and (23) can be readily carried out to obtain the trajectories.

Now it is noted that n is imaginary when $f^2 - \alpha^2 < 0$. In this case the solution of (19) must be written in the form

$$v = c_1 \cos int + c_2 \sin int, \tag{24}$$

where n is now real but cos int = cosh nt and sin int = sinh nt, thus $v = c_1 \cosh nt + ic_2 \sinh nt$ and, evaluating the constants,

$$v = v_0 \cosh nt + \frac{(\alpha^2 y_0 - fu_0)}{n} \sinh nt.$$

The equations for the velocity are therefore

$$\begin{aligned} u &= \frac{1}{f}\Big[\alpha^2 y - nv_0 \sinh nt - (\alpha^2 y_0 - fu_0) \sinh nt \\ v &= v_0 \cosh nt + \frac{(\alpha^2 y_0 - fu_0)}{n} \sinh nt. \end{aligned} \tag{25}$$

Equations (25) can be readily integrated to give trajectories for the above case when $\alpha^2 < f^2$. When $\alpha = f$ we have the special case

$$\begin{aligned} u &= \frac{\alpha^2}{f}(y - y_0) + u_0 \\ v &= v_0 + f\left(\frac{\alpha^2 y_0}{f - u_0}\right) t. \end{aligned} \tag{26}$$

Let us now consider a case where both pressure gradient and Coriolis parameter vary with latitude and put $f = 2\Omega y/a$. Then,

$$\begin{aligned} \frac{du}{dt} - \frac{2\Omega y}{a} v &= 0 \\ \frac{dv}{dt} + \frac{2\Omega y}{a} u &= \alpha^2 y. \end{aligned} \tag{27}$$

Then

$$\begin{aligned} u &= \frac{\Omega}{a}(y^2 - y_0^2) + u_0 \\ v^2 &= v_0^2 - \frac{\Omega^2}{a^2}(y^2 - y_0^2) + \frac{2\Omega u_0}{a}(y^2 - y_0^2) + \alpha^2(y^2 - y_0^2). \end{aligned} \tag{28}$$

Equation (28) is similar to (16) with the exception of the last term under the radical on the right-hand side.

If we consider air parcels starting from rest we have

$$v = \left[\alpha^2(y^2 - y_0^2) - \frac{\Omega^2}{a^2}(y^2 - y_0^2)^2\right]^{1/2}.$$

The peak of the trajectory is reached when $v = 0$, when

$$\frac{\Omega^2(y^2 - y_0^2)}{\alpha^2 a^2} = 1 \qquad \text{or} \qquad y^2 - y_0^2 = \frac{a^2\alpha^2}{\Omega^2}.$$

For air starting from rest from the equator the peak of the trajectory is

$$y = \frac{a\alpha}{\Omega} \tag{29}$$

Equation (29) may be compared with (18). It can be confirmed that a value of α^2 corresponding to a pressure gradient acceleration of 0.016 cm sec^{-2} at 5 degrees north latitude will give a value of y when the velocity is along the isobars that are close to that given by (18) for a constant pressure-gradient acceleration of 0.016 cm sec^{-2}.

It is of interest to study the case of a meridional pressure gradient invariant with latitude but changing with time. We then have

$$\frac{du}{dt} - fv = 0$$
$$\frac{dv}{dt} + fu = Bt \tag{30}$$

when $Bt = P_y$, the solutions of which are

$$u = u_0 \cos ft + \left(v_0 - \frac{B}{f^2}\right) \sin ft + \frac{Bt}{f}$$
$$v = \left(v_0 - \frac{B}{f^2}\right) \cos ft - u_0 \sin ft + \frac{B}{f^2} \tag{31}$$

and

$$x = \frac{u_0}{f} \sin ft + \frac{1}{f}\left(v_0 - \frac{B}{f^2}\right)(1 - \cos ft) + \frac{Bt^2}{2} + x_0$$
$$y = \frac{1}{f}\left(v_0 - \frac{B}{f^2}\right) \sin ft + \frac{u_0}{f}(\cos ft - 1) + \frac{Bt}{f^2} + y_0. \tag{32}$$

B is a proportionality constant that determines the time taken for the gradient to reach specified values P_y.

4. Special Coordinates

(a) Polar Coordinates. The equations of motion described in the previous section have been restricted to cases where the isobars are straight during the period of the integration. In order for a complete spatial representation of an actual synoptic pressure field to be included in a Cartesian coordinate trajectory model it would be necessary to devise a grid of points and assign numerical values of the pressure-gradient accelerations P_x and P_y in (1) to each square. Before doing this, however, it will be useful to look analytically at special cases of curved isobars. An example of such cases can be illustrated by systems of concentric, equally spaced isobars. There are two cases to consider, the anticyclonic and the cyclonic systems.

In studying such pressure fields it will be convenient to use a polar coordinate system in which equations (1) may be restated in the form

$$\ddot{r} - r\dot{\theta}^2 - fr\dot{\theta} = P_r$$
$$r\ddot{\theta} + 2\dot{r}\dot{\theta} + f\dot{r} = 0 \tag{33}$$

where r is measured outward from the origin, which is located at the center of the pressure system, and θ is measured in a counterclockwise direction about the origin.

If f is assumed to be a constant, integration of the second relation of (33) gives

$$r^2\dot{\theta} + \frac{fr^2}{2} = \text{const.} \tag{34}$$

If we assume first that a parcel of air starts from rest from the center of an anticyclone $r_0 = 0$, $\dot{\theta} = 0$ at $t = 0$, then the constant in (34) is zero, and $\dot{\theta} = -f/2$; substituting in the first relation of (33)

$$\ddot{r} + \frac{f^2 r}{4} - P_r = 0. \tag{35}$$

The solution of (35) for the initial conditions stated above is

$$r = \frac{4P_r}{f^2}\left(1 - \cos\frac{ft}{2}\right), \tag{36}$$

which is the equation of a cardioid.

If the parcel starts from rest from some distance r_0 from the center, (34) becomes

$$\dot{\theta} = \frac{f}{2}\left(\frac{r_0{}^2}{r^2} - 1\right). \tag{37}$$

Substituting the above expression into the first equation in (33) we have

$$\ddot{r} - \frac{f^2 r_0{}^4}{4r^3} + \frac{f^2 r}{4} - P_r = 0. \tag{38}$$

Multiplying through (38) by $\dot{r}$ gives

$$\dot{r}\ddot{r} - \frac{f^2 r_0{}^4 \dot{r}}{4r^3} + \frac{f^2 r\dot{r}}{4} - P_r\dot{r} = 0,$$

which may be readily integrated to give

$$\dot{r}^2 + \frac{f^2 r_0{}^4}{4r^2} + \frac{f^2 r^2}{4} - 2P_r r = \text{const.}$$

If the parcel starts from rest $\dot{r}_0 = 0$, and $r = r$ at $t = 0$, and const $= f^2 r_0{}^2/2 - 2P_r r_0$, and

$$\dot{r}^2 = 2P_r(r - r_0) - \frac{f^2}{4r^2}(r^2 - r_0{}^2)^2. \tag{39}$$

We may first examine (39) for the special case when $\dot{r} = 0$, that is, at the peak of the trajectory. We have

$$r^3 + r^2\left(r_0 - \frac{8P_r}{f^2}\right) - r r_0{}^2 - r_0{}^3 = 0. \tag{40}$$

Equation (40) may be solved for different values of the pressure gradient and Coriolis parameter. The solution for r will give the peak of the trajectory measured from the center of the anticyclone for cases in which the air parcels start from various distances r_0 from the center. This is shown in Figure 3. The peak distances at the outermost points on the trajectories for low latitudes, particularly for cases when the parcels start from the center of the anticyclone, are very far from the origin and quite unrealistic since the Coriolis parameter has been held constant. Thus, if a parcel started from the center of an anticyclone centered at 5 degrees of latitude and the Coriolis parameter was held constant at that latitude the parcel would reach 8,000 kilometers from the center before it started to return to its point of origin. Of course, the Coriolis parameter would in nature vary along the trajectory so that the parcels that left along a meridian extending northward from the center would reach the peak of their trajectories at much smaller radial distances.

Equation (39) may be solved and the radial velocities found at different radial distances from the center. $\dot{\theta}$ may also be computed from (34), and parcel trajectories constructed for cases when the parcels do not start from rest but possess initial speeds of $\dot{r}_0$ and $\dot{\theta}_0$. The allocation of a value to $\dot{r}_0$ will not complicate the equations. The allocation of a value to $\dot{\theta}_0$ will, however, turn the integration of Equation (38) into a long

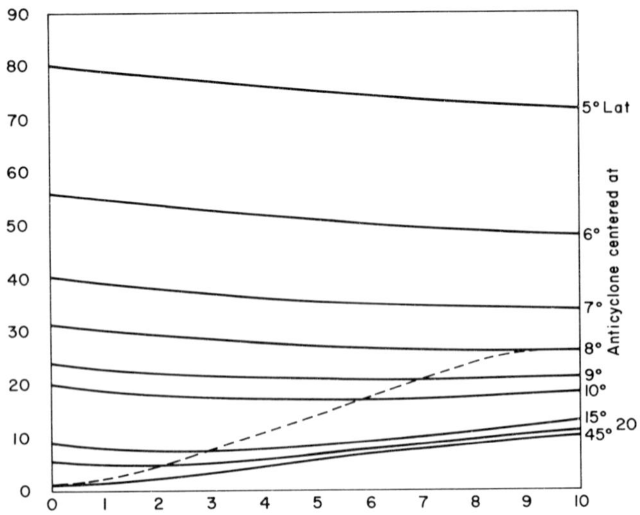

FIGURE 3. Distance from center of anticyclone of trajectory peak (left vertical axis; km × 10^2) for air parcels initially located various distances from the center of the anticyclone (horizontal axis; distances in degrees of latitude). Horizontal pressure gradient 0.016 cm sec^{-2}.

expression involving many terms. Since even then, the Coriolis parameter is assumed constant during the motion, and no friction has been introduced, there is little practical value in exploring such mathematical solutions further (*34*).

(b) Spherical Coordinates. Particle trajectories may also be computed from the solutions of the equations of motion expressed in spherical coordinates. Such a coordinate system automatically includes the variation of the Coriolis parameter in the analytical solutions since $f = 2\Omega \sin \phi$ is one of the coordinates used. For simplicity we will transform Equation (1) into the spherical coordinate form

$$\ddot{\lambda} \cos \phi - 2(\dot{\lambda} + \Omega)\, \dot{\phi} \sin \phi = 0$$
$$\ddot{\phi} + \dot{\lambda} \cos \phi \sin \phi\, (\dot{\lambda} + 2\Omega) = \frac{P_\phi}{a}. \tag{41}$$

The first of the above equations may be integrated to the form

$$\cos^2 \phi\, (\dot{\lambda} + \Omega) = \text{const.} \tag{42}$$

We assume initially that $\phi = \phi_0$, $\dot{\phi} = \dot{\phi}_0$, $\dot{\lambda} = \dot{\lambda}_0$, when $t = 0$. Then, (42) becomes

$$\cos^2 \phi\, (\dot{\lambda} + \Omega) - \cos^2 \phi_0\, (\dot{\lambda}_0 + \Omega) = 0. \tag{43}$$

Now substituting for $\dot{\lambda} + \Omega$ in the second relation in (41) we have,

since $$\ddot{\phi} + \sin \phi \cos \phi\, (\dot{\lambda} + \Omega - \Omega)\, (\dot{\lambda} + 2\Omega) = \frac{P_\phi}{a},$$

$$\ddot{\phi} + \sin\phi\cos\phi\,[(\dot{\lambda}_0 + \Omega)\cos^2\phi_0\sec^2\phi - \Omega]\,[(\dot{\lambda}_0 + \Omega)\cos^2\phi_0\sec^2\phi + \Omega] = \frac{P_\phi}{a} \tag{44}$$

$$\ddot{\phi} + \sin\phi\cos\phi\,[(\dot{\lambda}_0 + \Omega)^2\cos^4\phi_0\sec^4\phi - \Omega^2] = \frac{P_\phi}{a}.$$

Multiplying through by $\dot{\phi}$ and integrating give

$$\dot{\phi}^2 = \dot{\phi}_0{}^2 - \Omega^2(\cos^2\phi - \cos^2\phi_0) + \frac{2P_\phi}{a}(\phi - \phi_0) + (\dot{\lambda}_0 + \Omega)^2\cos^2\phi_0\,(1 - \cos^2\phi_0\sec^2\phi). \tag{45}$$

The meridional velocity can be computed from (45) at any latitude for a parcel originating its trajectory from latitude ϕ_0. Similarly the zonal velocity can be computed from (43). If the velocities are computed for small intervals of latitude, say 1 degree, the resulting trajectories may be constructed.

5. Trajectory Computations

The mathematical solutions of (1) in the various forms discussed in sections 1 to 4 are of mainly academic interest since the use of an electronic computer enables solutions of (6) and (7) to be obtained numerically. For this purpose it will be convenient to restate (6) and (7) in the form

$$\begin{aligned} u_n &= \left(u_{n-1} - \frac{P_y}{f}\right)\cos f\,\Delta t + \left(v_{n-1} + \frac{P_x}{f}\right)\sin f\,\Delta t + \frac{P_y}{f} \\ v_n &= \left(v_{n-1} + \frac{P_x}{f}\right)\cos f\,\Delta t - \left(u_{n-1} - \frac{P_y}{f}\right)\sin f\,\Delta t - \frac{P_x}{f} \end{aligned} \tag{46}$$

$$\begin{aligned} x_n &= \frac{1}{f}\left[P_y\,\Delta t + \left(u_{n-1} - \frac{P_y}{f}\right)\sin f\,\Delta t + \left(v_{n-1} + \frac{P_x}{f}\right)(1 - \cos f\,\Delta t)\right] + x_{n-1} \\ y_n &= \frac{1}{f}\left[-P_x\,\Delta t + \left(v_{n-1} + \frac{P_x}{f}\right)\sin f\,\Delta t - \left(u_{n-1} - \frac{P_y}{f}\right)(1 - \cos f\,\Delta t)\right] + y_{n-1} \end{aligned} \tag{47}$$

where $u_{n-1}, v_{n-1}, x_{n-1}, y_{n-1}$ are the component velocities and coordinates at the end of the preceding time step, t is the time interval selected (one hour), u_n, v_n, x_n, y_n are the component velocities and coordinates at the end of the nth time step.

B. INTRODUCTION OF FRICTION

The velocities computed using (46) become unrealistically large during the course of the integration due to the continued acceleration. This acceleration occasionally results in parcels apparently meeting one another at relatively early stages of the motion and gives rise to difficulties in the computation and interpretation of the divergence field.

1. Coefficient of Surface Resistance

The cases considered in this report all refer to low-level flow over the ocean. Such flow is subject to frictional resistance and a more realistic model should therefore include a damping factor.

It is not considered essential to specify the detailed mathematical form expressing the change of stress with height. Such terms in the equation of motion would take the form

$$\begin{aligned} \frac{du}{dt} - fv &= -\frac{1}{\rho}\frac{\partial P}{\partial x} + \frac{1}{\rho}\frac{\partial \tau_x}{\partial z} \\ \frac{dv}{dt} + fu &= -\frac{1}{\rho}\frac{\partial P}{\partial y} + \frac{1}{\rho}\frac{\partial \tau_y}{\partial z} \end{aligned} \tag{48}$$

where τ_x, τ_y are the eddy-shearing stresses along the x and y axes respectively. Rather it is considered simpler to return to the damping factors used by Guldberg and Mohn (*29*) and many other workers who incorporate a surface resistance term that is proportional to the wind speed. Thus,

$$\begin{aligned} \frac{du}{dt} - fv &= -\frac{1}{\rho}\frac{\partial P}{\partial x} - ku \\ \frac{dv}{dt} + fu &= -\frac{1}{\rho}\frac{\partial P}{\partial y} - kv \end{aligned} \tag{49}$$

where k is a resistance coefficient. The resistance terms will refer to some layer of air bounded by the sea surface. The actual thickness of the layer considered would depend on the magnitude of k used. A discussion of the applicability of a damping factor that is proportional to the wind speed and opposed to it and of the choice of suitable numerical values for k will be offered after a solution of (49) has been derived.

We may express the first relation of (49) in terms of v

$$v = \frac{1}{f}\left(\frac{du}{dt} + ku - P_x\right) \tag{50}$$

and $\dot{v} = 1/f(\ddot{u} + k\dot{u})$, assuming P_x to be held constant. Substituting $\dot{v}$ in the second relation of (49) we have

$$\ddot{u} + 2k\dot{u} + (k^2 + f^2)u - P_x k - fP_y = 0.$$

We will let $P_x + P_y = C$ and $(f^2 + k^2)u - C = q$. Then

$$\ddot{q} + 2k\dot{q} + (f^2 + k^2)q = 0$$

or

$$\ddot{q} + 2a\dot{q} + bq = 0. \tag{51}$$

An algebraic equation of type

$$r^2 + 2ar + b = 0$$

has roots r_1 and r_2 where $r_1 = \alpha + i\beta$ and $r_2 = \alpha - i\beta$.
Then $r^2 + 2ar + b = r^2 - r(r_1 + r_2) + r_1 r_2$, equating coefficients

$$2a = -r_1 - r_2$$

or

$$a = \frac{-r_1 - r_2}{2} = \alpha$$

$$b = r_1 r_2 = \alpha^2 + \beta^2$$

or

$$\beta^2 = b^2 - a^2.$$

The solution of (51) is of form

$$q = c_1 f_1(t) + c_2 f_2(t)$$

where c_1 and c_2 are constants and

$$f_1(t) = e^{\alpha t} \cos \beta t$$
$$f_2(t) = e^{\alpha t} \sin \beta t.$$

Thus

$$q = c_1 e^{\alpha t} \cos \beta t + c_2 e^{\alpha t} \sin \beta t$$
$$q = c_1 e^{-at} \cos (b^2 - a^2)^{1/2} t + c_2 e^{-at} \sin (b^2 - a^2)^{1/2} t$$

or

$$q = c_1 e^{-kt} \cos ft + c_2 e^{-kt} \sin ft$$

and

$$u = \frac{e^{-kt}}{f^2 + k^2}(c_1 \cos ft + c_2 \sin ft) + \frac{c}{f^2 + k^2} \tag{52}$$

also, from (45)

$$v = \frac{e^{-kt}}{f^2 + k^2}(c_2 \cos ft - c_1 \sin ft) + \frac{kc}{f(f^2 + k^2)}. \tag{53}$$

To evaluate the constants c_1 and c_2 we let $u = u_0$, $v = v_0$ at $t = 0$ as for the frictionless case.

Then

$$c_1 = u_0(f^2 + k^2) - P_x k - fP_y$$
$$c_2 = v_0(f^2 + k^2) - \frac{kc}{f} + \frac{P_x(f^2 - k^2)}{f}$$

and (52) and (53) become

$$\begin{aligned} u = {} & \frac{e^{-kt}}{f^2 + k^2}\Bigg\{[u_0(f^2 + k^2) - P_x k - fP_y] \cos ft \\ & + \left[v_0(f^2 + k^2) - \frac{k}{f}(P_x k + fP_y) + \frac{P_x}{f}(f^2 + k^2)\right] \sin ft\Bigg\} \\ & + \frac{P_x k + fP_y}{f^2 + k^2} \\ v = {} & \frac{e^{-kt}}{f^2 + k^2}\Bigg\{\left[v_0(f^2 + k^2) - \frac{k}{f}(P_x k + fP_y) + \frac{P_x}{f}(f^2 + k^2)\right] \cos ft \\ & - [u_0(f^2 + k^2) - P_x k - fP_y] \sin ft\Bigg\} + \frac{k(P_x k + fP_y)}{f(f^2 + k^2)} - \frac{P_x}{f}. \end{aligned} \tag{54}$$

Equations (54) may now be integrated to give the coordinates of the trajectories as a function of time.

Let $\dfrac{P_x k + fP_y}{f^2 + k^2} = M$ and $\dfrac{kM - P_x}{f} = N.$

Then,

$$\begin{aligned} \dot{x} &= e^{-kt} \cos ft(u_0 - M) + e^{-kt} \sin ft(v_0 - N) + M \\ \dot{y} &= e^{-kt} \cos ft(v_0 - N) - e^{-kt} \sin ft(u_0 - M) + N \end{aligned} \tag{55}$$

integrating

$$\begin{aligned} x = {} & \frac{(u_0 - M)}{f^2 + k^2} e^{-kt}(f \sin ft - k \cos ft) - \frac{(v_0 - N)}{f^2 + k^2} e^{-kt}(k \sin ft \\ & + f \cos ft) + Mt + \frac{k(u_0 - M) + f(v_0 - N)}{f^2 + k^2} + x_0 \end{aligned} \tag{56}$$

$$y = \frac{(v_0 - N)}{f^2 + k^2} e^{-kt}(f \sin ft - k \cos ft) + \frac{(u_0 - M)}{f^2 + k^2} e^{-kt}(k \sin ft$$
$$+ f \cos ft) + Nt + \frac{k(v_0 - N) - f(u_0 - M)}{f^2 + k^2} + y_0.$$

It is seen that when $k = 0$ equations (55) and (56) reduce to (6) and (7) respectively.

The formulation of friction deserves some comment. Firstly, it follows the original formulation of Guldberg and Mohn, and doubtless is suggested by familiar examples of mechanical damping. Indeed, Rayleigh (*53*) suggested the same form of damping in his dissipation function, where the latter is just twice the dissipation, and leads to a considerable simplification in Lagrange's equation.

However, when applied to fluid motions a frictional force proportional to the speed implies laminar flow, a doubtful assumption in the earth's boundary layer. Nevertheless, it is clearly necessary to require some dissipation throughout, say, the lowest 1,000 meters of the atmosphere; the present formulation is the simplest one in which (48) can be integrated, and, at the same time, contain some essential physics. Moreover, Phillips (*54*) and Kao (*55*) have utilized this formulation in the former's well-known numerical experiment and in the latter's investigation of non-linear particle oscillations in an atmospheric long wave. Again, the motivation was a desire to incorporate some dissipation, not to investigate the mechanics of turbulence.

The justification for the present formulation of friction is then frankly empirical: does it lead to sensible speeds? Further, and perhaps more importantly, with friction opposing the motion, rather than acting at some angle as implied in the Ekman theory, the present formulation applies if the wind does not turn with height. Judging from Gray's (*56*) mean statistics from the tropical oceans, this assumption is reasonably good. At worst, the small changes in direction through the lowest kilometer (about 10°) are known with insufficient precision to warrant additional refinements. Finally, Swinbank (*57*) has shown that the frictional force is linearly dependent on the wind speed, providing: (1) horizontal kinetic energy is conserved, and (2) the shearing stress and (reversed) wind direction are quasi-parallel, the latter being deduced observationally from the "Leipzig Profile."

Balanced frictional flow is often derived from either (48) or (49) by assuming that the motion is not accelerated. Equation (49) then takes the form

$$\begin{aligned} -fv + ku &= P_x \\ fu + kv &= P_y. \end{aligned} \tag{57}$$

For the sake of convenience we may orient the x axis along the isobars. Then,

$$\begin{aligned} -fv + ku &= 0 \\ fu + kv &= P_y \end{aligned}$$

and

$$\tan \alpha = \frac{v}{u} = \frac{k}{f} \tag{58}$$

where α is the angle between the isobars and the wind direction.

Equation (58) may be applied to synoptic or climatological charts of the wind and pressure fields to estimate k. If synoptic charts are used the cases chosen should be for balanced motion when the flow is straight and when there is no isallobaric wind that would contribute to the ageostrophic component. If climatological charts are used care should be taken also to select areas where the mean flow is not accelerated and where the isobars are straight. The computation of the surface resistance coefficient should be carried out for an atmosphere in which the lapse rate does not depart too widely from the dry adiabatic. If the lower layers are unstable there will be downward transport of horizontal momentum from levels where the air is moving at geostrophic speed and consequently the cross isobaric angle in Equation (58) will be smaller than for the more stable case.

A number of estimates have been made of the mean angle of cross isobaric flow as a function of the latitude and of the lapse rate over the sea. Individual values in unstable conditions in the temperate zone westerlies may be negative (*58*). Mean values of α (*59*) vary from about 10 degrees in the temperate zone westerlies to about 40 degrees at latitude 5 degrees. Such a range of α gives corresponding range of k from about 1.0×10^{-5} to 3.0×10^{-5}. Values within this range were obtained from Equation (58) for the Indian Ocean and Arabian Sea when localities were chosen where the mean flow, as depicted on climatological charts, was not accelerated.

A second independent method of estimating the coefficient of surface resistance is to use Equation (48). For balanced frictional flow we have, assuming that the isobars are oriented along the x axis,

$$\begin{aligned} -fv - \frac{1}{\rho}\frac{\partial \tau_x}{\partial z} &= 0 \\ fu - \frac{1}{\rho}\frac{\partial \tau_y}{\partial z} &= P_y. \end{aligned} \qquad (59)$$

Let us put $k\mathbf{V} = -\frac{1}{\rho}\frac{\partial \boldsymbol{\tau}}{\partial z}$

where $\mathbf{V}$ is the vector wind and $\boldsymbol{\tau}$ is the vector stress that is directed along the motion. The vector form of Equation (48) is

$$\frac{d\mathbf{V}}{dt} \times \mathbf{K} + f(\mathbf{V} - \mathbf{V}_g) = -\frac{1}{\rho}\frac{\partial \boldsymbol{\tau}}{\partial z} \qquad (60)$$

where $\mathbf{K}$ is the unit vector normal to the x, y plane, and $\mathbf{V}g$ is the vector geostrophic wind, which for balance flow becomes

$$f(\mathbf{V} - \mathbf{V}_g) = -\frac{1}{\rho}\frac{\partial \boldsymbol{\tau}}{\partial z}. \qquad (61)$$

Equation (61) may be integrated between the surface and the top of the frictional layer. Thus

$$\boldsymbol{\tau}_0 = -\int_0^H \rho f(\mathbf{V} - \mathbf{V}_g)\, dz \qquad (62)$$

where $\boldsymbol{\tau}_0$ is the stress at the surface, assuming that there is no stress at the top of the friction layer. The expression on the right-hand side of Equation (62) has been evaluated over the ocean from pilot balloon ascents. If the profile of the ageostrophic wind is plotted against height the area bounded by the z axis and the wind profile curve extending from the surface to the top of the frictional layer will be proportional to the difference in stress between the base and top of the frictional layer.

Experimental values so obtained give stresses of the order of 1 dyne cm^{-2} (*59*). If the stress at the surface is large it is likely that the top of the friction layer will be higher than when the stress at the surface is small. Thus $1/\rho\ \partial\tau/\partial z$ may be fairly conservative when evaluated throughout the frictional layer.

If we suppose that $\Delta\tau$ is 1 dyne cm^{-2} between the surface and 500 meters, ρ is 1.2×10^{-3} grams cm^{-3} and V is 10 m sec^{-1}, we obtain $k = 1.7 \times 10^{-5}$, a value which compares favorably with those obtained from (*58*).

2. Divergence and Vertical Motion

The families of trajectories that can be constructed by the Lagrangian solutions of the equations of motion are of considerable interest in understanding air motion in the tropics. In particular they offer a means of computing the divergence field which, in turn, by invoking the equation of continuity, yields a field of vertical motion.

$$\text{div}\ \mathbf{V}\ dt = \frac{dA}{A} = \log\frac{A_2}{A_1} = \log\frac{A_n}{A_{n-1}} \tag{63}$$

where div $\mathbf{V}$ is the horizontal divergence of the wind and A_n/A_{n-1} is the ratio of the contraction or dilation of the area bounded by three or more parcels during the interval of time Δt_n. If the zonal pressure gradient is neglected ($A = 0$), Equation (63) becomes

$$\text{div}\ \mathbf{V}\ dt = \log\frac{\Delta y_n}{\Delta y_{n-1}} \tag{64}$$

where

$$\frac{\Delta y_n}{\Delta y_{n-1}} = \frac{y_{k,n} - y_{(k+1),n}}{y_{k,(n-1)}, - y_{(k+1),(n-1)}}, \tag{65}$$

k designates the number of the parcel that starts from some initial latitude and n is the number of the time step. Thus, starting at the initial latitude the numerator expresses the difference in the y coordinate separating two parcels 1 and 2 at $t = 0$. The denominator expresses the distance separating the same two parcels after a time interval of one hour. When $P_x = 0$, or when the pressure gradient is only a function of y and not of x, the trajectories of the parcels are unaffected by the meridian of origin. They will, in effect, duplicate one another. Thus the computed field of divergence will be given by Equation (64) for parcels which begin their trajectories from grid point locations at different latitudes along the same meridian.

The divergence is given by the ratios of the one-dimensional distances separating the parcels for the case under discussion since any parcels originating to the east or west of a given parcel which is being followed will at all times remain at the same distance east or west of its neighboring parcel. Thus the areas will at all times be parallelograms and the ratios of the areas of these parallelograms will be the ratios of their altitudes, which are given by Equation (65).

It is important to state that the technique described computes the divergence between a field of parcels which are initially located within selected boundary lines at the same instant in time. Divergence is not computed between any parcels which subsequently enter the area. This restraint becomes less important as the boundaries of the selected area are expanded. It becomes nonexistent when the initial field of points covers the whole globe.

The trajectories which have been discussed so far have been computed under the assumption that the motion was horizontal. This was implicitly assumed in the formulation of the equations of motion in which hydrostatic equilibrium is implied, and the effect of the vertical component of the Coriolis parameter in generating a vertical acceleration has been neglected. However, divergence of the horizontal motion has been interpreted in the sense that downward vertical motion is occurring and conversely convergence of the horizontal motion has been interpreted in the sense that upward vertical motion is taking place. Now the trajectories have been computed for parcels of air of unit mass and they therefore refer to the movement of the centers of mass of these parcels. If the parcels undergo vertical motion one way or the other they will be stretched or compressed vertically. If the stretching of a parcel of air is large the element of unit mass, which was originally, say, cubic in shape, would be transformed to an elongated filament. The top of such a filament might extend to a level where the pressure gradient differed from the value at the original lower level of the center of mass of the parcel and this could result in a discrepancy between a computed and an actual trajectory. On the other hand it might well be convenient and reasonable to assume that the air parcels which are being followed possess the function of bounding an area which is divergent or convergent and that although vertical motion occurs within that area the bounding parcels themselves do not partake of such vertical motion.

The vertical motion of the area bounded by the parcels whose trajectories are computed from the equations as discussed may be obtained from the relation

$$\frac{1}{A}\frac{dA}{dt} + \frac{1}{z}\frac{dz}{dt} = 0 \tag{66}$$

Equation (66) states that the total divergence is zero. Horizontal divergence is therefore compensated by vertical convergence and vice-versa.

Thus $$\frac{1}{z}\frac{dz}{dt} = -\operatorname{div} \mathbf{V}$$

and $$\log\frac{z_1 - z_L}{z_2 - z_L} = \operatorname{div} \mathbf{V}\,\Delta t = \log\frac{\Delta y_2}{\Delta y_1} = \log\frac{\Delta z_1}{\Delta z_2} \tag{67}$$

where z_L is the lower reference level of the air parcel and z_1, z_2 are the upper boundaries of the parcel at successive time intervals. Parcels chosen in the frictional layer may have their lower boundary at the surface. Then

$$z_L = 0 \text{ and } \log\frac{z_1}{z_2} = \log\frac{\Delta y_2}{\Delta y_1}.$$

For the more general case for the frictional layer

$$\log\frac{z_{n-1}}{z_n} = \log\frac{\Delta y_n}{\Delta y_{n-1}} \tag{68}$$

where the term on the right-hand side is defined by Equation (65). Equation (68) then becomes

$$z_n = z_{n-1}\frac{(\Delta y_{n-1})}{\Delta y_n}. \tag{69}$$

In order to compute values of the z coordinate from Equation (69) we may choose an initial level z_0 at $t = 0$ which is representative of the upper boundary of the air parcel to be followed. Such initial levels may be 1, 10, 100 meters, et cetera, as desired. Now,

$$\frac{dw}{dz} = -\operatorname{div} \mathbf{V}. \tag{70}$$

The first value of the vertical velocity to be computed would be at $t = 1$ hour when the first divergence value is known. The height of the parcel at $t = 1$ hour is

$$z_1 = z_0 \frac{\Delta y_0}{\Delta y_1}. \tag{71}$$

At the boundary layer between the atmosphere and the surface $w = w_L = 0$ and

$$z = z_L = 0. \quad \text{At } t = 1 \text{ hour}$$

$$(w_1 - w_L)_{t=1} = (z_L - z_1)_{t=1} \operatorname{div} \mathbf{V}$$

or

$$w_1 = -z_1 \operatorname{div} \mathbf{V} = -z_1 \log \frac{\Delta y_1}{\Delta y_0}. \tag{72}$$

For subsequent times, from Equation (66)

$$w_n = z_n \log \frac{\Delta y_{n-1}}{\Delta y_n}. \tag{73}$$

IV. *Some Computations of Air Motions in Low Latitudes*

In this section we offer a brief résumé of previous investigations, followed by five case studies of near surface trajectories, taken from both the Indian and Pacific oceans. The five examples include some no-drift aircraft flights over the Indian Ocean in June and July of 1963; in these we attempt to compute a sample steady-state trajectory and infer some crude features of friction. This is followed by three climatological examples—the Marshall Islands, the Indian Ocean, and the Gilbert-Ellice Islands—where we attempt to compute some steady-state trajectories, and the accompanying horizontal divergence. We compare the latter results with the available climatological distributions of cloud and rainfall.

We offer a few computations of theoretical interest, including double near-equatorial troughs symmetrically placed with respect to the equator; a pressure gradient along the equator; and some computed trajectories in both circular and elliptical isobaric systems. In all the above, the interpretation is suggestive rather than definitive.

Lastly, through an appeal to the divergence equation, we offer some interpretation of the four climatological cases.

A. PREVIOUS WORK

Serious investigation of pressure and wind in low latitudes has a reasonably long though very broken history. Guldberg and Mohn (*29*) first published a (graphical) solution based on the horizontal equations of motion through a specification of the pressure field; though rather crude, their trajectories did depict some major features of the summer circulation over the Indian Ocean. However, the experimental evidence was meager, and, not surprisingly, their results did not elicit general conviction among a wider audience of dynamical meteorologists.

After Shaw and Gold offered some convincing proof of the validity of the gradient and geostrophic wind formulations, Shaw was of the opinion that if any pressure-wind relationship was valid for low latitudes it could only be the gradient wind. Such an assertion necessarily remained speculative without convincing observational evidence; and this was not available until World War II.

Grimes (*60*), appealing to particle dynamics, offered some theoretical speculations concerning wind and pressure, assuming, among other things, non-divergent flow.

With the dramatic expansion of observations in the tropics, particularly the Pacific, during World War II, Treloar, Gibbs, and, later,

Crossley (*61*) continued the examination of air motions in low latitudes. The latter two investigators again appealed to particle dynamics and attempted to incorporate friction (Gibbs) and the likely effects of accelerations close to the equator. Gibbs employed the Guldberg-Mohn formulation of friction and attempted to evaluate the frictional constant, k, through a direct appeal to measured pressure gradients in low latitudes. These and other developments were summarized by Grimes (*62*). Later, Gordon and Shaw (*52*), again appealing to particle dynamics, attempted to derive a low-latitude general-circulation model.

With the exception of the early attempt by Guldberg and Mohn (*29*) to compute the field of motion directly from the observed pressure field, only two recent attempts appear to have used this dynamical approach: one by Wyrtki (*63*), valid at the surface over the oceanic tropics, and another by Johnson and Mörth (*64*) working in tropical Africa.

Wyrtki attempted to compute the field of (horizontal) motion directly from the pressure field using the same formulations embodied in Equation (57) but assuming that the horizontal accelerations vanished; with this assumption, a simple computation shows that the scalar wind speed and direction are given by

$$v = \left[\frac{P_x^{\,2} + P_y^{\,2}}{k^2 + f^2}\right]^{1/2}, \tan \varphi = \frac{k}{f}. \tag{74}$$

After specifying three simple pressure fields (... "equidistant isobars parallel to the equator ... isobars crossing the equator under 45°, and ... a pressure trough at 8 degrees north" ...) the speed and direction of the wind is computed from Equation (74), assuming "that the movements are stationary" (*63*, p. 14).

Johnson and Mörth (*64*), (*65*) working in tropical Africa, have summarized their experience in the form of several synoptic models relating wind and pressure in the lower troposphere. Later, Mörth (*66*) made minor amendments to two of the four basic models, but he has not apparently altered his interpretive framework.

Briefly, the above investigators described several models of wind and pressure, together with inferred divergence patterns, that have, in their judgment, provided some insight into forecasting over East Africa. Their synoptic results of particular flow patterns and observed weather are encouraging, and apparently have led to improved forecasts. Nevertheless, their results warrant comment on two grounds: (*a*) their ability to analyze the pressure field with requisite detail in equatorial latitudes, and (*b*) their appeal to particle dynamics as an interpretive framework.

Firstly, it is difficult to measure the surface pressure accurately in low latitudes, even on isolated atolls; further, to measure pressure-heights with conventional radiosondes from the same atolls to heights above 700 mb is more difficult still. To add to this a continental station network, wherein the surface elevation itself is uncertain, to say nothing of using radiosondes of various manufacture, is to increase the overall uncertainty by a considerable amount. A discussion of these factors would have been a welcome addition to the synoptic charts presented by Johnson and Mörth, and would have contributed in some measure to understanding their subjective treatment of some of the observations.

To fix ideas, for the 200 mb surface to be accurately analyzed at height intervals of 40 gpm, as the above authors have done, the mean virtual temperature of the air column between 850 and 200 mb must be known to within 1°C. Similarly, the 500 mb surface could be accurately analyzed at intervals of 20 gpm if the mean virtual temperature were known to within about 1.3°C. In low latitudes, the latter requirement can probably be met if the network uses one make of radiosonde; the former is marginal. Besides, computed pressure-heights over a network where surface elevations themselves are not known with confidence cannot be accepted without reservation. Unfortunately, these nagging doubts have not been removed by Thompson (*67*) in an otherwise extensive work on the climatology of Africa. In fact, even analysis of mean pressure-heights obtained from a sea-level network of uniform instrumentation entails serious limitations. In this regard, Ballif's (*68*) experience in the Marshall Islands is instructive: mean pressure-height "values consistent with those at the surface were obtained at the 850 and 700 mb levels; above this, no significant pattern could be drawn" (p. 28).

Secondly, Johnson and Mörth (*65*) attempted to furnish a theoretical interpretation of their empirical results within the framework of non-viscous particle dynamics. They presented four pressure distributions, with attendant wind fields—the equatorial "Duct" (low pressure on the equator, high to the north and south), the "Drift" (high to the north, low to the south of the equator), the "Bridge" (high pressure on the equator, low to the north and south), and a "Composite Model." Numerically computed trajectories were presented for the "Duct," the "Drift," and a "trajectory illustrating deflection into westerly downgradient flow"; the "Bridge" and "Composite Model" were schematic only. Lastly, in one case (the "Drift"), an analytical solution was compared with the numerical one.

After specifying the pressure field, the authors unfortunately failed to demonstrate that their initial states were, in some way, preferred ones. This is not a trivial point; indeed, it can be demonstrated quite easily (*69*) that with a constant pressure-gradient force and Coriolis parameter, solutions of the horizontal, inviscid equations of motion admit of three different kinds of trajectories depending, respectively, on whether the magnitude of the geostrophic velocity is greater than, equal to, or smaller than that of the initial velocity of geostrophic deviation. That is, whether

$$|u_g| \gtreqless |u_0 - u_g|. \tag{75}$$

In particular, the three cases are wave-shaped, cycloidal, and a series of loops, respectively.

The fact that Johnson and Mörth have incorporated a variable Coriolis parameter, and that the above criteria are for a constant one, does not alter the basic objection; for example, the above analysis could easily be applied to their Figure 7.1.23—where the parcel amplitude is about 2 degrees of latitude—obtaining a very different result using different initial speeds. Thus, Johnson and Mörth's assertion— "... for a given pressure gradient in the ideal drift, there is a critical latitude equatorwards of which geostrophic flow is only in neutral stability and, for *all* subgeostrophic velocities the air will flow southwards across the equator. As one proceeds polewards of the critical latitude the flow is stable for increasingly subgeostrophic winds,"—is not, in the general case, a valid conclusion (*65*, p. 64). In fact, a "stable" trajectory can be achieved at any latitude provided the initial state is managed properly.

Though it is, in general, true that such a "critical latitude" exists under the conditions noted, it is also true that all parcels would be "stable" if the initial speeds were supergeostrophic. Moreover, it is instructive to realize how sensitive this critical latitude is; for example,

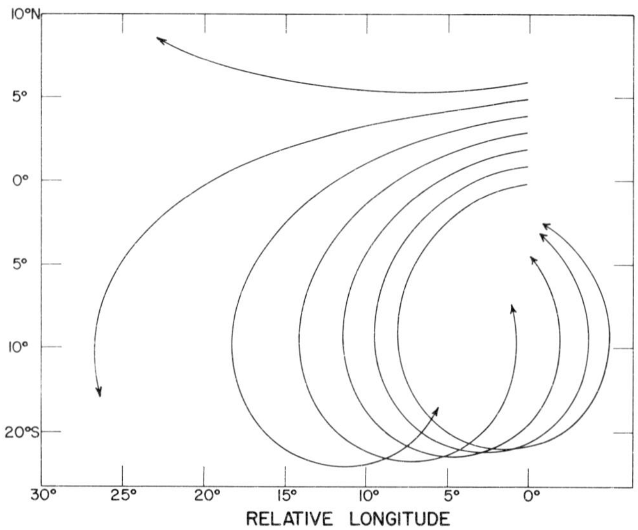

FIGURE 4. Seven sample trajectories computed from Johnson and Mörth's (*65*) Figure 7.1.24. They give initial conditions for only the parcel at 5°N ($U_0 = -9.80$, $V_0 = 1.50$ m sec^{-1}); the remaining initial speeds were assigned by arbitrarily adding 10 cm sec^{-1} to the u-components equatorward of 5°N, and subtracting the same amount from the v-component over the same distance.

with a constant meridional gradient of 1.4 mb/5° latitude (23.0×10^{-3} cm sec^{-2}), and an initial easterly speed of -13.6×10^2 cm sec^{-1} at 5°N, and -6.8×10^2 cm sec^{-1} at 10°N, the parcel at 5°N will cross the equator, while the one at 10°N will not. On the other hand, with a gradient of 1.6 mb/5° latitude (25.0×10^{-3} cm sec^{-2}) and the same initial speeds, both parcels now cross the equator. The initial speeds are both subgeostrophic: in the first case by 4.50×10^2 cm sec^{-1} and 2.26×10^2 cm sec^{-1} respectively; in the second case by 6.07×10^2 cm sec^{-1}, and 3.05×10^2 cm sec^{-1}, respectively. Thus, a difference of 79.0 cm sec^{-1} in the two geostrophic deviations at 10°N is sufficient to determine whether or not the parcel at 10°N would or would not cross the equator! Such distinctions are vital in determining the character of the divergence; the latter is very sensitive to the initial state (*70*). The change of the divergence with respect to small changes in the initial speed is inversely proportional to both the geostrophic deviation ($u_0 - u_g$) and the change of initial speed with respect to initial latitude. Is it to be inferred that the necessary discrimination of initial speeds has been obtained from the author's synoptic experience?

Again, consider the 7 trajectories in Figure 4 computed from Equations (6 and 7) with $k = 0$, and adopting the pressure gradient (12.7×10^{-3} cm sec^{-2}) and initial speeds ($u_0 = -980$, $v_0 = -150$ cm sec^{-1}) provided by Johnson and Mörth (*65*) in their Figure 7.1.24; the other initial speeds were determined by arbitrarily adding 10 cm sec^{-1} to the u-component equatorward of 5°N, and subtracting the same amount from the v-component over the same distance. The trajectory of the parcel leaving 5°N agrees well with that obtained by the authors. On the other hand, the parcel leaving 6°N travels northward performing "a stable oscillation," while the rest cross the equator. Under these conditions the "critical latitude" is somewhere between 5° and 6°N, whereas in the example discussed previously it was north of 10°N. Since the number of air parcels actually crossing the equator is proportional to the longitudinal extent of the divergence zone to the south, the precise location of the "critical latitude" is of considerable synoptic interest. Again, this can be assured if the initial states are carefully selected.

By way of contrasting and comparing particle trajectories for the "Drift" and "Duct," Johnson and Mörth presented "a comparison between the trajectories of the two particles which initially have identical velocities in identical gradients at 5°N.... In one case the pressure gradient is constant, in the other the meridional profile is bowl-shaped as appropriate to a "Duct." The "Duct" case is stable; the drift is not. This result is due to the decrease of the "Duct," and in very low latitudes air which has only a slightly subgeostrophic speed will attempt to cross the equator..." (*65*, p. 64). These comparisons are presented in their Figure 7.1.24, while the "bowl-shaped" pressure profiles are depicted in their Figure 7.1.20. Unfortunately, it is not possible to provide a comparison similar to that in Figure 4 since the authors have not identified which "bowl-shaped" profile they have used. Nevertheless, their "stable" trajectory could easily be altered to something dramatically different by changing the initial speeds by as little as a few tenths of a meter per second.

Lastly, similar comments could be directed at the results presented in the authors' Figure 7.1.27 ("Trajectory illustrating deflection into westerly downgradient flow"); for example, if the particle were directed either toward the south or due north the time and location of the particle approaching the equator would be vastly different. Again, is the chosen initial state a preferred one drawn from either synoptic or climatological experience?

B. CASE STUDIES

1. No-Drift Aircraft Flights

During the International Indian Ocean Expedition a number of "no-drift" heading flights were made by aircraft of the United States Weather Bureau Research Flight Facility in the summer of 1963. Such flights offered a unique opportunity to examine a number of features of atmospheric flow in low latitudes. It was observed from synoptic charts that approximate steady-state conditions existed for the monsoon current during the period of the flights. The aircraft flight tracks could thus be assumed to duplicate the trajectories of the monsoon current.

In the following sections some equations are given from which we may calculate a number of atmospheric parameters useful for studying the flow of the monsoon current. These equations are presented in normal coordinates that are ideal for sampling the flow from a Lagrangian observational system that is forced to travel continuously in a direction tangent to the fluid motion (or directly opposed to it).

(*a*) *The Cross-Contour Velocity Factor.* The extent to which the wind crosses the isobars is a problem that is difficult to determine, particularly in the free atmosphere above the surface frictional level. No-drift flights afford an occasion to investigate this aspect of the wind flow.

In normal coordinates the equations of motion for horizontal frictionless flow may be written

$$\begin{aligned} \frac{dV_s}{dt} - fV_n &= -g\frac{\partial z}{\partial s} \\ \frac{dV_n}{dt} + fV_s &= -g\frac{\partial z}{\partial n} \end{aligned} \tag{76}$$

where V_s, V_n are the velocities along and normal to the flow and $\partial z/\partial s$, $\partial z/\partial n$ are the contour gradients along and normal to the flow, respectively. Now during a no-drift flight the aircraft flies into or down the wind continuously. $V_n = 0$ but dV_n/dt may not be zero, and (76) becomes for steady-state conditions

$$\begin{aligned} V_s\frac{\partial V_s}{\partial s} &= -g\frac{\partial z}{\partial s} \\ V_s{}^2\frac{\partial \theta}{\partial s} + fV_s &= -g\frac{\partial z}{\partial n} \end{aligned} \tag{77}$$

where $\partial\theta/\partial s$, the turning of the wind, is also the curvature of the trajectory.

The value of the contour gradient along the direction of motion may be computed, together with the change in height of the air parcels as they move along the trajectory. The contour gradient normal to the flow may also be computed from the second relation of (77).

This will be recognized as the gradient wind equation. However, since the wind is not necessarily a gradient wind the terms in the first relation will not be zero. This equation describes the nongradient part of the wind. The second relation expresses the component of the actual wind that blows along the isobars, that is the part of the wind that is gradient.

The contour gradient along the track, $\partial z/\partial s$, was obtained from the equation

$$dh = \frac{\partial z}{\partial s}ds + \frac{\partial z}{\partial p}dp \tag{78}$$

where dh is the change in height measured by the aircraft radar altimeter and dp is the change in barometric pressure, both measured along an increment ds of the flight path. Then

$$\frac{\partial z}{\partial s} = \frac{dh}{ds} + \frac{dP}{\rho g\, ds}. \tag{79}$$

The first flight, which is examined here, left Bombay on the morning of 26 June 1963 and headed into the wind. At 11.1°N, 55.7°E it headed toward Aden away from the no-drift track. On the following morning, 27 June, the aircraft returned to its no-drift heading at 10.2°N, 54.7°E and continued to 1.6°N, 47.3°E where it turned off toward Nairobi. The level of the flight was maintained at about 500 meters above the sea.

The return flight left Nairobi on 1 July 1963. The no-drift segment of this flight ran from 1.64°N, 46.79°E to 9.86°N, 53.48°E. At the latter position the aircraft headed for Aden. The following morning, 2 July 1963, the aircraft returned to Bombay, reaching the no-drift segment at 10.89°N, 54.09°E and concluding the no-drift segment at 18.68°N, 67.83°E before landing at Bombay.

The values, measured at 15-minute intervals, of latitude, longitude, barometric pressure, radar altitude, temperature, and wind speed and direction along the no-drift track are listed in Table 1 for the two legs

TABLE 1. Position, height (feet), pressure (mb), temperature (°K), and wind (ft/sec) averaged over 1-minute intervals every 15 minutes along the Bombay-Aden-Nairobi no-drift flight.

Time	Latitude (degrees)	Longitude (degrees)	Radar Altitude (feet)	Pressure (mb)	Temperature (degrees A)	Wind Speed (ft/sec)	Wind Direction (degrees)
0300	19.035	72.494	1624	952.58	296	18.6	271
0315	18.889	71.623	1628	953.15	296	20.0	279
0330	18.941	70.762	1645	952.01	296	25.3	284
0345	19.097	69.909	1634	952.45	296	21.5	269
0400	19.117	69.028	1621	952.88	296	23.6	256
0415	19.022	68.169	1636	952.34	294	26.3	258
0430	18.828	67.359	1584	953.49	295	28.7	258
0445	18.622	66.530	1602	953.62	295	26.5	258
0500	18.405	65.712	1610	952.98	295	35.5	250
0515	18.194	64.896	1610	953.14	295	30.4	246
0530	17.891	64.119	1610	952.95	295	35.5	244
0545	17.561	63.387	1612	952.70	295	40.0	243
0600	17.211	62.656	1619	952.42	295	42.2	235
0615	16.838	61.963	1621	952.82	295	47.8	234
0630	16.405	61.300	1628	953.05	295	42.2	235
0645	15.932	60.669	1626	953.11	295	41.0	235
0700	15.480	60.011	1665	952.41	294	40.5	235
0715	15.026	59.365	1625	953.11	294	39.3	235
0730	14.536	58.759	1649	952.27	294	47.3	225
0745	14.030	58.180	1636	952.97	294	51.9	223
0800	13.484	57.658	1670	952.45	295	47.3	226
0815	12.898	57.142	1668	953.45	294	43.0	218
0830	12.305	56.656	1682	952.06	294	47.3	217
0845	11.706	56.172	1689	951.86	294	52.6	224
0900	11.104	55.700	1698	951.55	294	55.7	223
0900	10.207	54.699	1629	952.63	294	55.7	219
0915	9.646	54.251	1644	952.54	294	55.6	222
0930	9.073	53.785	1653	952.22	294	54.0	227
0945	8.522	53.296	1634	952.97	294	50.8	228
1000	8.015	52.750	1641	952.69	293	50.7	230
1015	7.500	52.227	1645	952.64	294	48.5	223
1030	6.957	51.698	1661	952.37	293	43.9	227
1045	6.420	51.174	1666	952.69	294	41.7	229
1100	5.890	50.650	1673	952.21	293	42.2	232
1115	5.409	50.068	1662	952.25	293	43.8	234
1130	4.916	49.486	1697	951.84	293	50.7	224
1145	4.394	48.979	1700	951.98	293	47.5	234
1200	3.862	48.446	1677	952.46	294	37.2	213
1215	3.266	47.962	1706	951.54	292	34.8	215
1230	2.610	47.530	1701	951.97	293	32.0	194
1245	1.841	47.351	1718	951.36	293	29.4	186

TABLE 2. Position, height (feet), pressure (mb), temperature (°K), and wind speed (ft/sec) averaged over 1-minute intervals every 15 minutes along the Nairobi-Aden-Bombay no-drift flight.

Time	Latitude (degrees)	Longitude (degrees)	Radar Altitude (feet)	Pressure (mb)	Temperature (degrees A)	Wind Speed (ft/sec)	Wind Direction (degrees)
0800	1.644	46.786	1780	95.178	292	22.0	182
0815	2.486	47.086	1762	95.201	293	23.6	206
0830	3.299	47.521	1746	95.213	292	28.7	213
0845	3.992	48.145	1762	95.121	292	37.2	234
0900	4.694	48.828	1683	95.319	292	43.9	224
0915	5.419	49.466	1670	95.305	293	49.0	223
0930	6.120	50.140	1653	95.306	292	40.5	223
0945	6.802	50.833	1622	95.349	293	47.3	219
1000	7.565	51.483	1618	95.268	293	52.4	219
1015	8.322	52.142	1582	95.319	294	57.4	222
1030	9.087	52.812	1568	95.257	294	62.5	221
1045	9.864	53.484	1515	95.292	294	64.2	218
0800	10.893	54.089	1691	94.961	294	62.5	211
0815	11.640	54.683	1678	94.986	294	62.5	223
0830	12.331	55.403	1505	95.371	294	72.6	227
0845	13.084	56.184	1505	95.286	294	65.9	224
0900	13.801	56.946	1501	95.286	294	59.1	229
0915	14.503	57.711	1481	95.283	294	59.1	233
0930	15.157	58.523	1446	95.322	295	57.4	234
0945	15.793	59.332	1441	95.294	295	50.7	235
1000	16.400	60.138	1426	95.312	295	47.3	240
1015	16.991	60.965	1436	95.297	296	38.8	244
1030	17.498	61.839	1439	95.287	295	37.2	250
1045	17.883	62.758	1419	95.299	296	35.5	249
1100	18.221	63.728	1411	95.257	296	40.5	257
1115	18.506	64.738	1379	95.355	296	35.5	260
1130	18.671	65.754	1375	95.336	296	33.8	272
1145	18.704	66.778	1368	95.351	296	28.7	273
1200	18.683	67.833	1285	95.426	296	54.0	268

TABLE 3. ESSA research flight facility instrumentation and error estimates for the DC-6A, Indian Ocean, 1963.

Parameter	Instrument	Range	Error	Time Constant
1. Static pressure	Pressure transducer	50–1,050 mb	0.05 mb	10 mb/sec
2. Altitude above ground level	Radar	0–50,000 ft	±25 ft	50 ft/sec
3. Wind direction	Doppler navigation	0–360°	Vector	36°/sec
4. Wind speed	Doppler navigation	0–240 knots	error: ±5 kts	10 knots/sec
5. Latitude	Doppler navigation	90°N or S	1% of change	
6. Longitude	Doppler navigation	180°E or W	1% of change	
7. Temperature	Resistance bridge, vortex thermometer	−60 to +40°C	±1°C	2°/sec

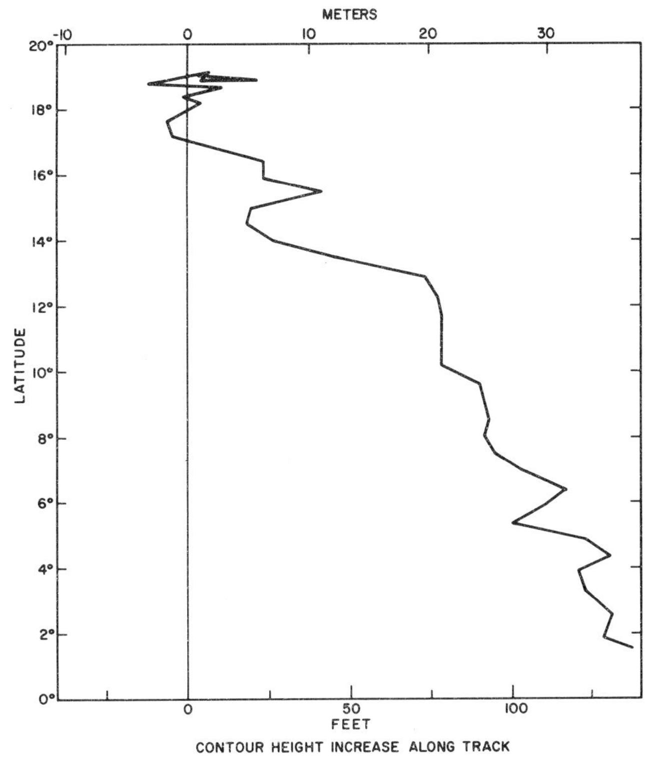

FIGURE 5. Profile of the increase of height of the constant-pressure surface along the flight track as a function of the latitude (Bombay to Nairobi).

of the outward flight, and in Table 2 for the two legs of the return flight. Each value is the mean of seven individual observations reported at 10-second intervals of flying time and centered about the middle value listed in Column 1. Each tabulated value is therefore an average over a minute of flying time or approximately 4 miles. Table 3 gives some details of the instrumentation and error estimates for the research aircraft used in the flights. The 1-minute averages should smooth errors due to the tolerance and response of the instrumentation.

The accumulated changes in height of the pressure surface along the flight tracks as computed from (79) are shown in Figures 5 and 6. With Bombay as the reference value, the contour surface on the outward flight rose a total of about 42 m, which is equivalent to a pressure rise of about 5 millibars. The monsoon current was therefore blowing down the pressure gradient on its journey from the equator to the latitude of Bombay. On the return flight the pressure gradient was steeper. The total decrease in height of the constant pressure surface at the flight level was about 159 m between the equator and Bombay.

The return flight in Figure 6 is shown in two parts since the aircraft reached the no-drift heading at a higher latitude on its way out of Aden

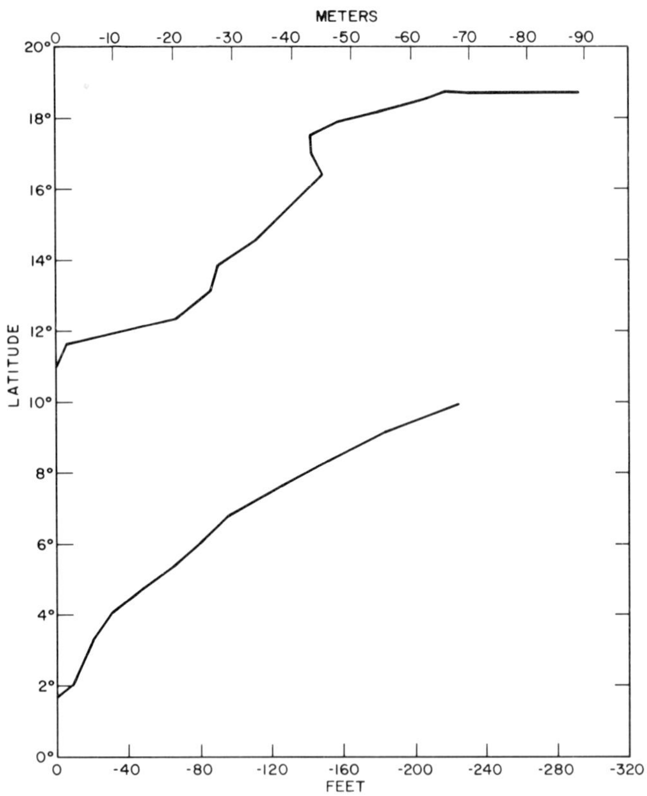

FIGURE 6. Profile of the decrease of height of the constant-pressure surface along the track as a function of the latitude (Nairobi-Bombay). The two legs Nairobi-Aden and Aden-Bombay are shown separately.

than it had left the no-drift heading on its way into Aden the previous day.

Figure 7 shows the mean sea-level pressure profile for the outward and return flights respectively. The profiles were computed from the thickness equation assuming a lapse rate of 9°C km^{-1}, estimated from dropsondes. The mean sea-level values are in good agreement with those reported by ships in the vicinity and plotted on synoptic maps for the periods of these flights (Figures 23 and 24). It is noted here again that the pressure gradient was considerably steeper on the return flight than on the outward flight. The diurnal variation of pressure is partly responsible for the discontinuity in the pressure profile at the latitude where the two flight legs join.

The gradient itself $\partial z/\partial s$ has been plotted for 15-minute intervals during the outward flight in Figure 8. In Figures 9 and 10 the contour gradients for both outward and return flights have been computed over time intervals of 30 minutes. This is equivalent to an approximate distance of 120 miles. There are wide fluctuations in the gradient over the 15-minute interval band, including belts over which the wind appears to be blowing up the gradient. This counter-gradient flow diminishes considerably when the gradient is averaged over 30-minute intervals of flying time. In Figures 8 to 10 a positive gradient signifies that the pressure surfaces slope downward from south to north.

(*b*) *Contour Gradient Perpendicular to the Wind Flow.* The contour gradient perpendicular to the wind flow, that is, to the flight track, may be obtained from the second relation of (77). Thus,

$$-\frac{\partial z}{\partial n} = \frac{V_s^2}{g}\frac{\partial \theta}{\partial s} + \frac{fV_s}{g}. \tag{80}$$

The resultant pressure gradient normal to the flow is seen to be given by the sum of the geostrophic and rotational components of the gradient. If the flow is cyclonic the latter two components are both acting to the left of the flow in the northern hemisphere. If the flow is anticyclonic the rotational component is negative and directed in the opposite direction to the geostrophic component. If the resultant gradient is zero the flow is inertial. If the resultant gradient is negative and directed to the right of the flow the flow is antibaric and dynamically unstable.

Figure 11 shows the component of the pressure gradient normal to the flow averaged over 15 minutes of flying time for the outward flight. This component also fluctuates considerably and the fluctuations decrease with increasing latitude. South of 5°N the flow becomes antibaric.

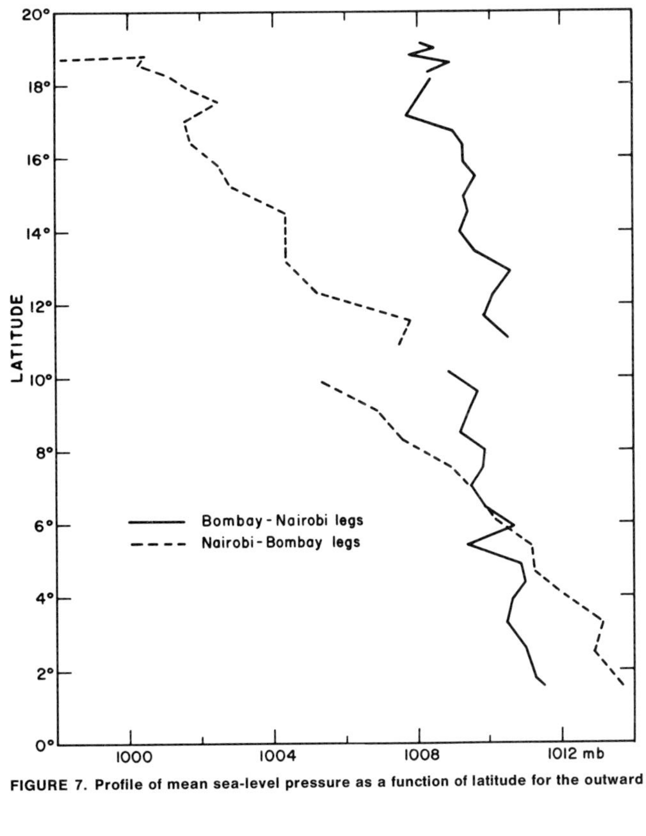

FIGURE 7. Profile of mean sea-level pressure as a function of latitude for the outward and return flights.

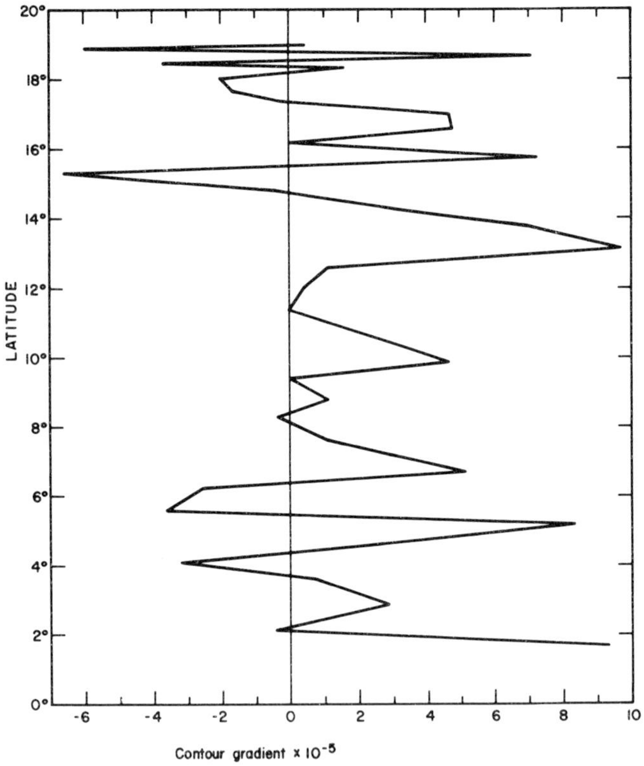

FIGURE 8. Profile of the contour gradient along the track as a function of the latitude averaged over 15-minute intervals of flying (outward flight). Positive values of the gradient mean that the pressure surface slopes downward from south to north.

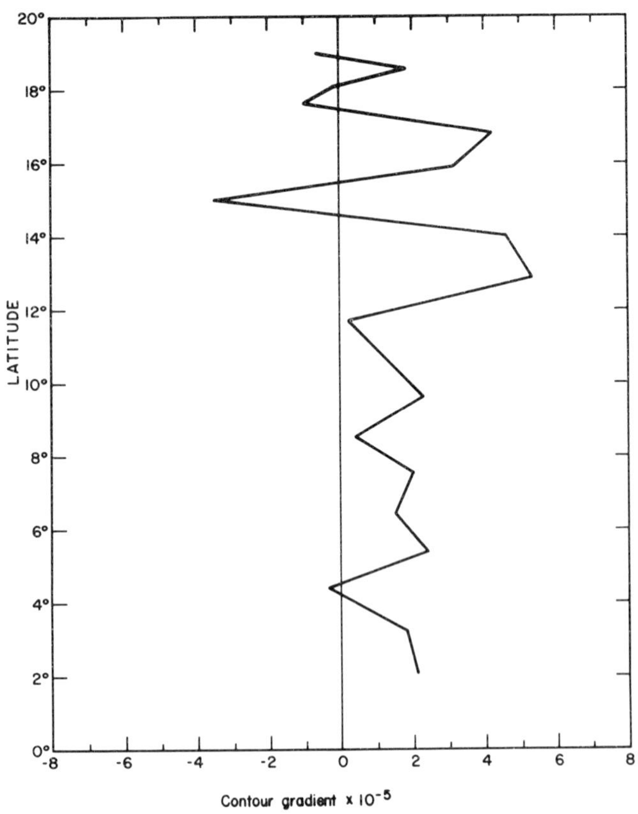

FIGURE 9. Profile of the contour gradient along the track as a function of the latitude averaged over 30-minute intervals of flying time (outward flight).

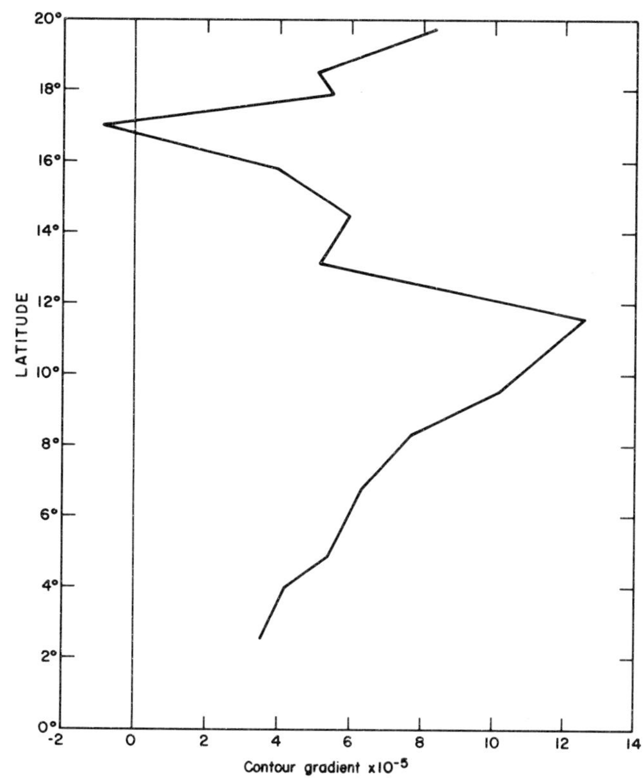

FIGURE 10. Profile of the contour gradient along the track as a function of the latitude averaged over 30 minutes of flying time (return flight).

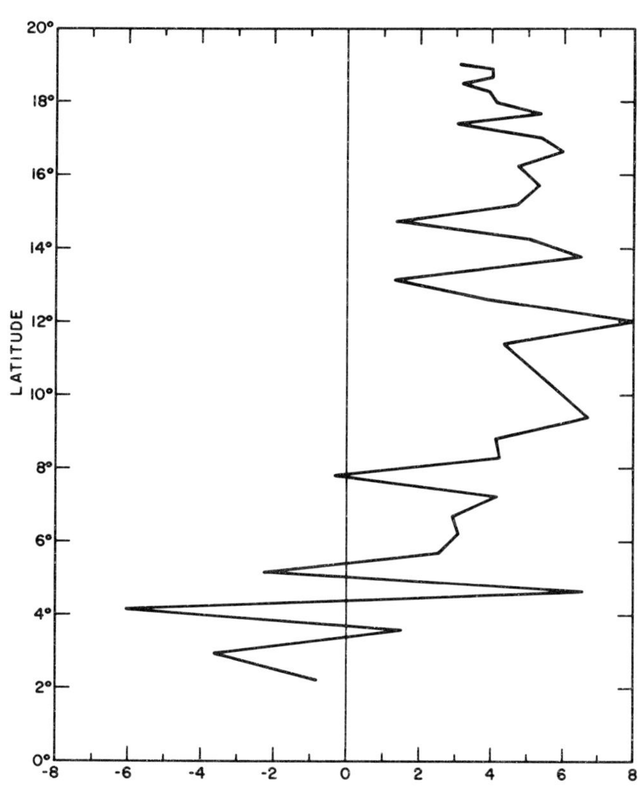

FIGURE 11. Contour gradient normal to the track as a function of latitude averaged over 15-minute intervals of flying time (outward flight).

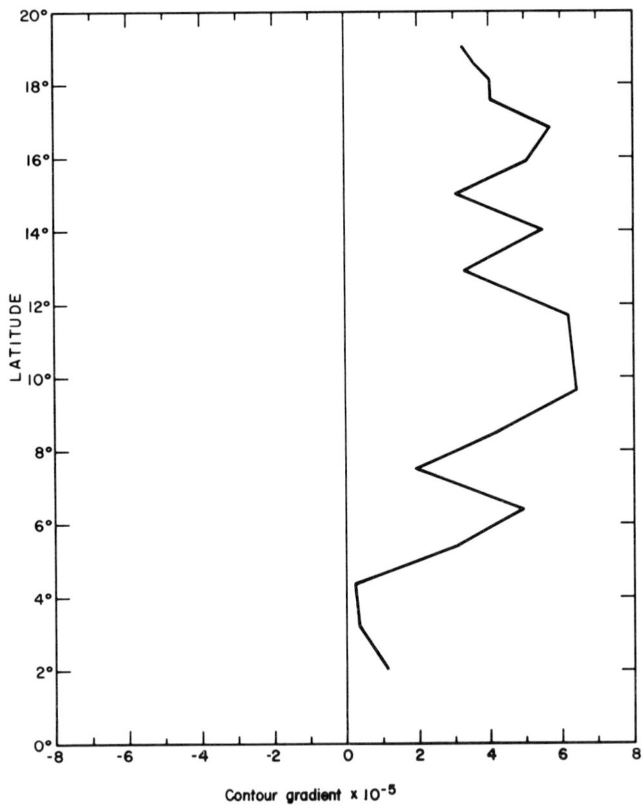

FIGURE 12. Contour gradient normal to the track as a function of latitude averaged over 30-minute intervals of flying time (outward flight).

In Figures 12 and 13 the gradient normal to the flow has been averaged over 30-minute intervals of flight time for both outward and return flights. The flow becomes almost inertial at 4°N latitude. A positive gradient signifies that the pressure surfaces slope downward to the left of the flow of the wind in the geostrophic sense.

The computed gradients along and perpendicular to the flow have been resolved into isobars in Figures 23 and 24. A comparison of the orientation of the isobars computed by (79) and (80) with the pressure field analyzed from composite actual weather maps for the periods of the flights shows good agreement.

(*c*) *Angle of Inclination of the Wind to the Isobars.* One of the most informative parameters for studying the interaction of the wind and pressure fields in low latitudes is the angle at which the wind blows across the isobars. If α represents this angle we have

$$\alpha = \tan^{-1} \frac{\dfrac{\partial z}{\partial s}}{\dfrac{\partial z}{\partial n}}. \qquad (81)$$

Figures 14 and 15 show profiles of the angle α plotted over 30-minute intervals of flying time for the outward and return flights respectively; absolute values of α greater than 90 degrees indicate antibaric flow. Positive values indicate that the wind is blowing from high to low pressure and vice-versa.

(*d*) *The Ageostrophic Wind.* The ageostrophic wind may be resolved into components along the flow and normal to the flow (68). These may be denoted by V_{ags} and V_{agn} respectively, where

$$V_{agn} = \frac{-g}{f}\frac{\partial z}{\partial s}, \text{ and } V_{ags} = \frac{V^2}{f}\frac{\partial \theta}{\partial s}$$
$$V_{ag} = \left[\frac{g^2}{f^2}\left(\frac{\partial z}{\partial s}\right)^2 + \frac{V^4}{f^2}\left(\frac{\partial \theta}{\partial s}\right)^2\right]^{1/2}. \tag{82}$$

Figure 16 shows profiles of the ageostrophic wind averaged over 30-minute intervals of flying time for the return flight. The departure of the wind from its geostrophic value is seen to increase with decreasing latitude in accordance with expectations.

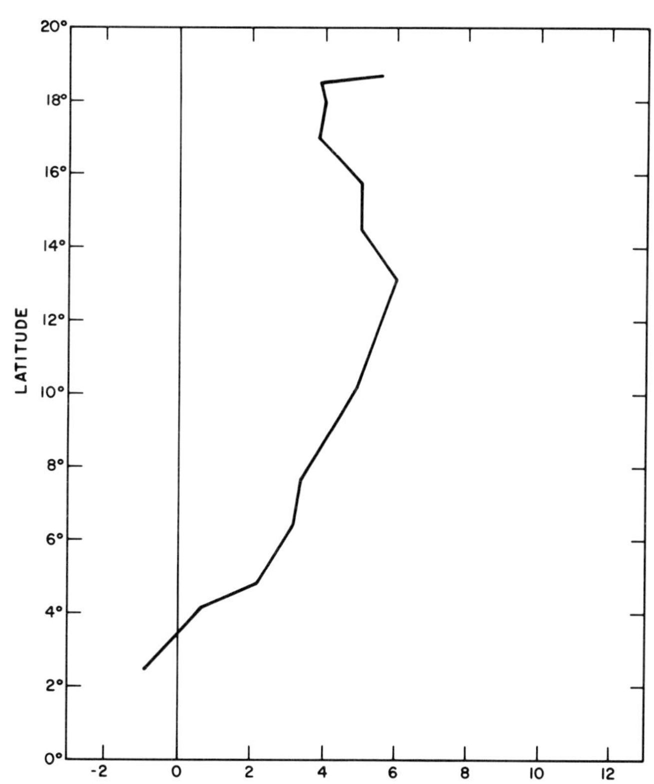

FIGURE 13. Profile of the contour gradient normal to the track as a function of the latitude averaged over 30 minutes of flying time (return flight).

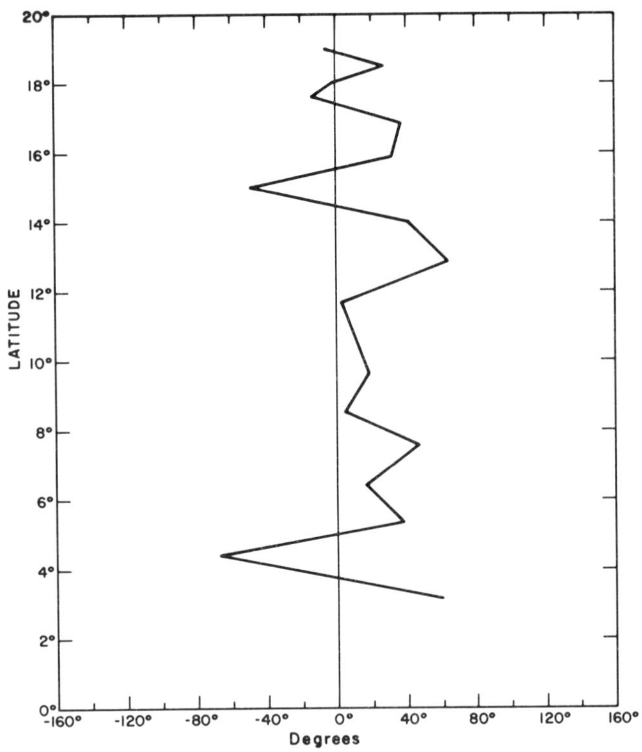

FIGURE 14. Profile of angle of inclination of wind to isobars as a function of latitude averaged over 30-minute intervals of flying time (outward flight).

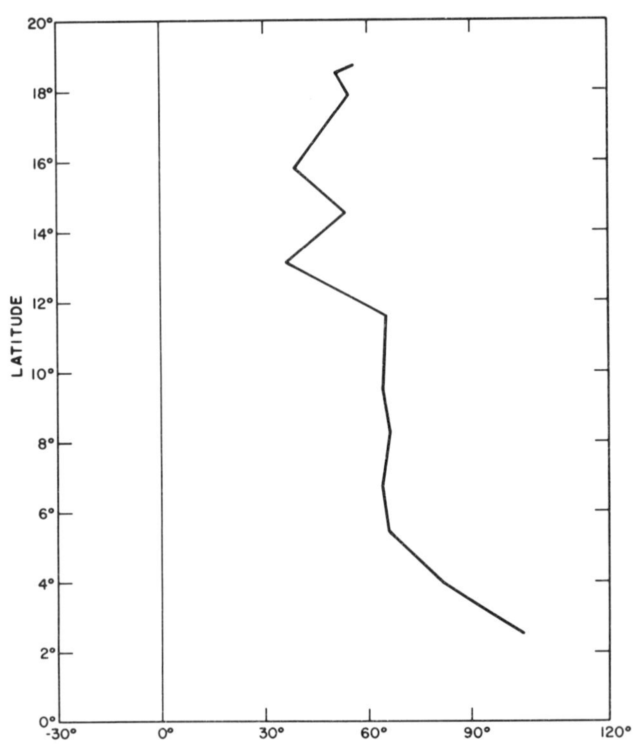

FIGURE 15. Profile of the inclination of the wind to the isobars as a function of latitude averaged over 30 minutes of flying time (return flight).

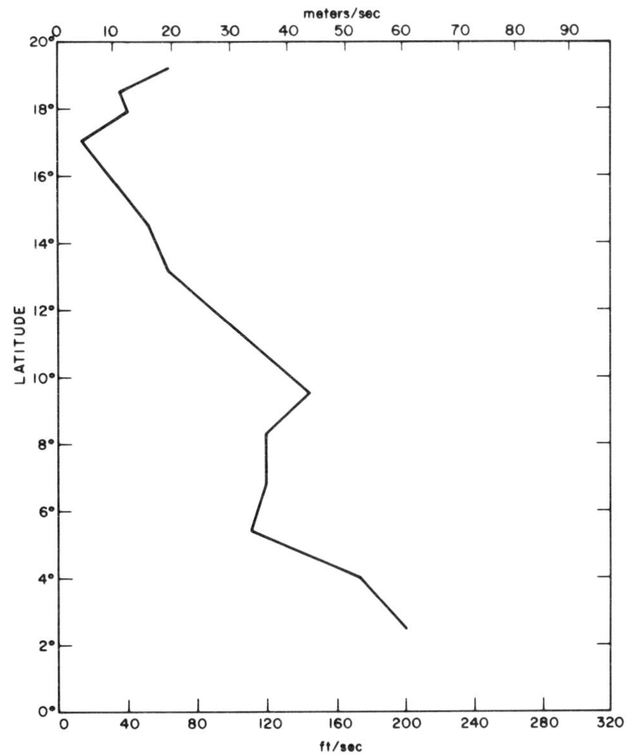

FIGURE 16. Profile of the ageostrophic wind as a function of latitude averaged over 30 minutes of flying time (return flight).

(*e*) *Divergence.* Divergence in normal coordinates is given by

$$\text{div } \mathbf{V} = \frac{\partial V}{\partial s} + V_s \frac{\partial \Psi}{\partial n} \tag{83}$$

where $\partial \Psi / \partial n$ is the change in direction of the trajectories as measured along the axis normal to the flow. When only one trajectory is available as in the case of the no-drift flight trajectories studied here one is forced to assume that the term $V_s \, \partial \Psi / \partial n$ is small compared to $\partial V / \partial s$ and may be neglected. The term $\partial V / \partial s$ has been plotted as a function of latitude for the outward flight in Figure 17. The divergence shows wide fluctuations but the profile indicates a gradual change from divergence at the lower latitudes to convergence north of 10°N.

(*f*) *Relative Vorticity.* Relative vorticity in normal coordinates is given by

$$\xi = V \frac{\partial \Psi}{\partial s} - \frac{\partial V_s}{\partial n}. \tag{84}$$

The first term represents the curvature vorticity. The second term expresses the shear vorticity. In computing relative vorticity following the flow along only one trajectory it is only the curvature component that can be included. Assuming then that the shear component is small

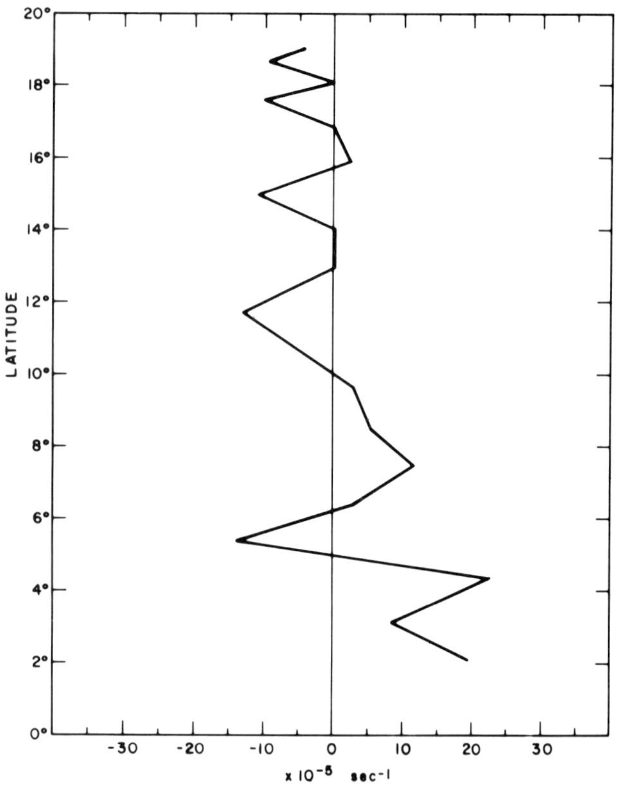

FIGURE 17. Profile of speed divergence (∂V/∂s) as a function of latitude averaged over 30 minutes of flying time (outward flight).

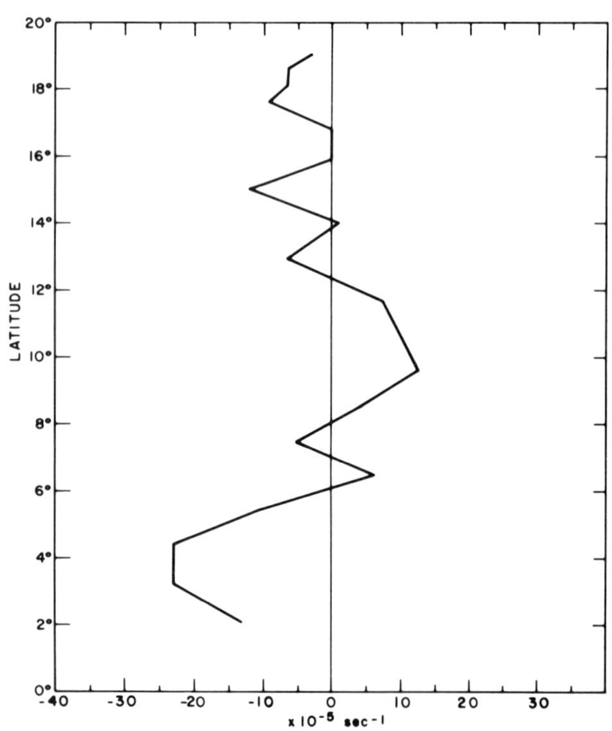

FIGURE 18. Profile of curvature vorticity as a function of latitude averaged over 30 minutes of flying time (outward flight).

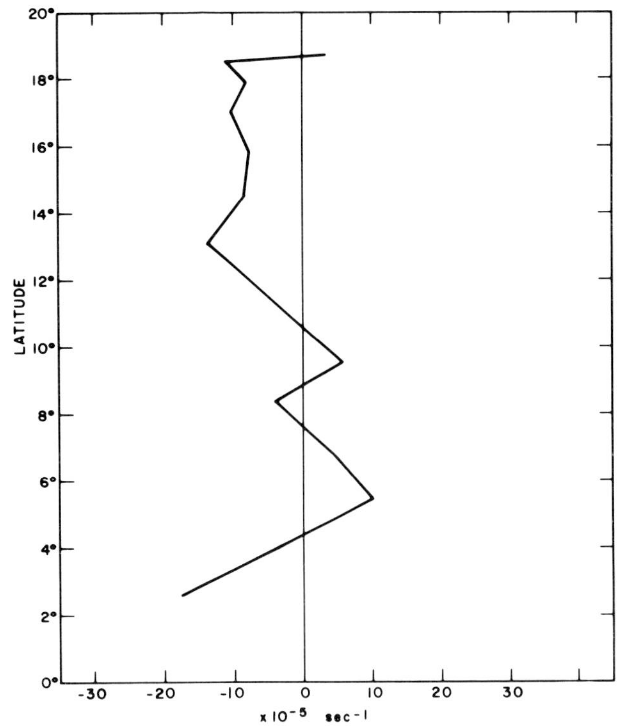

FIGURE 19. Profile of curvature vorticity as a function of latitude averaged over 30 minutes of flying time (return flight).

the relative vorticity of the flow along the track has been plotted as a function of latitude in Figures 18 and 19.

The relative vorticity, though fluctuating widely, appears to be negative or anticyclonic near the equator and also north of about 12°N. There is positive or cyclonic vorticity between the two negative belts. It is emphasized that the neglect of the shear vorticity may cause a pattern of total relative vorticity that differs from the true one.

Figure 20 shows the profile of absolute vorticity obtained by adding the Coriolis parameter to the profile for the return flight in Figure 19. It is of interest to note that the absolute vorticity becomes negative south of 4°N.

(*g*) *Coefficient of Frictional Resistance and Surface Stress.* If the frictional terms are added to the equations of motion in normal coordinates in the form given in (49) we have for a steady state

$$V_s \frac{\partial V_s}{\partial s} = -g \frac{\partial z}{\partial s} - kV_s$$
$$V_s^2 \frac{\partial \theta}{\partial s} + fV_s = -g \frac{\partial z}{\partial n}. \tag{85}$$

Thus,

$$k = -\frac{\partial V_s}{\partial s} - \frac{g}{V_s}\frac{\partial z}{\partial s} \tag{86}$$

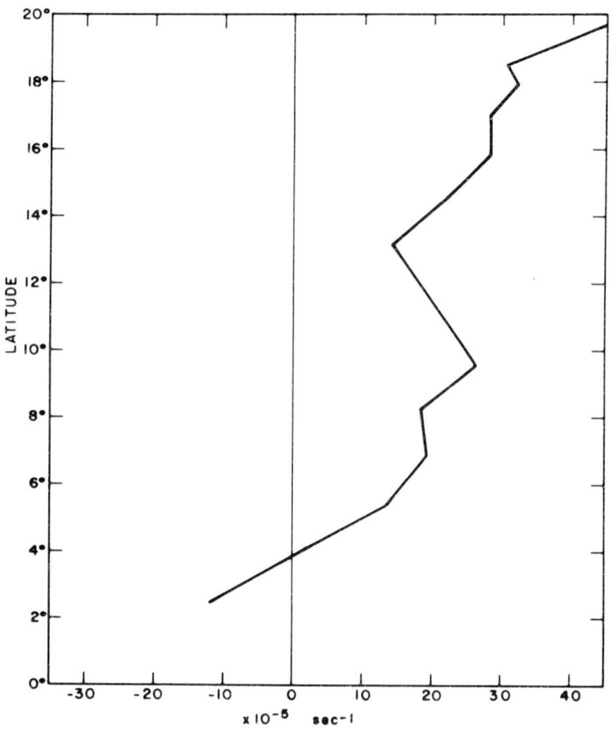

FIGURE 20. Profile of absolute vorticity as a function of latitude averaged over 30 minutes of flying time (return flight).

where a positive numerical value of k denotes that friction is opposing the motion. Values of k have been plotted as a function of latitude in Figures 21 and 22 for the outward and return flights respectively.

The frictional acceleration term in (85) is usually expressed as $-1/\rho\ \partial\tau_s/\partial z$ where τ_s is the stress that is directed along the flow. The numerical values of $-1/\rho\ \partial\tau_s/\partial z$ are given by the difference between the pressure-gradient acceleration and the actual acceleration along the flow. If it should be assumed that $-1/\rho\ \partial\tau_s/\partial z$ is constant from the flight level down to the surface and that τ_s is very small at the flight level, which is near the top of the frictional layer, we have $\tau_0 = B\rho\, dz$, where B is the evaluated quantity $-1/\rho\ \partial\tau_s/\partial z$ and τ_0 is the surface stress.

Values of τ_0 were computed in this way from the aircraft measurements of altimeter height, pressure, and temperature averaged over 1-minute intervals. The values of τ_0 then refer to segments of the trajectory separated by 15-minute intervals of flying time. The mean value of τ_0 for the two no-drift legs of the Bombay-Nairobi flight was 0.9 dynes/cm^2. The mean value of the coefficient of frictional resistance at flight level was 1.3×10^{-5} sec^{-1} for the outward flight and 3.5×10^{-5} sec^{-1} for the return flight. The higher value for the return flight might be due to a larger value of the exchange coefficient of momentum in the more turbulent flow associated with a larger pressure gradient.

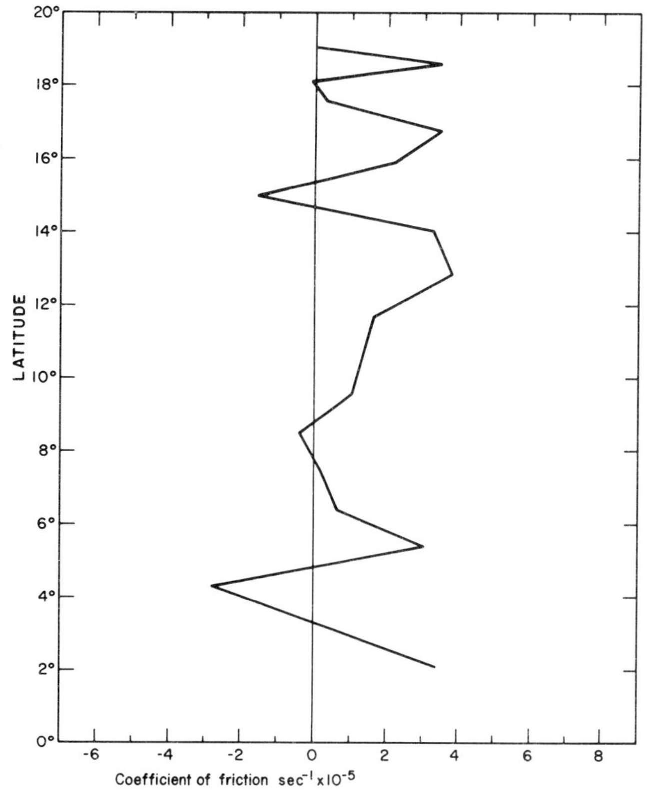

FIGURE 21. Profile of the coefficient of frictional resistance as a function of latitude averaged over 30 minutes of flying time (outward flight).

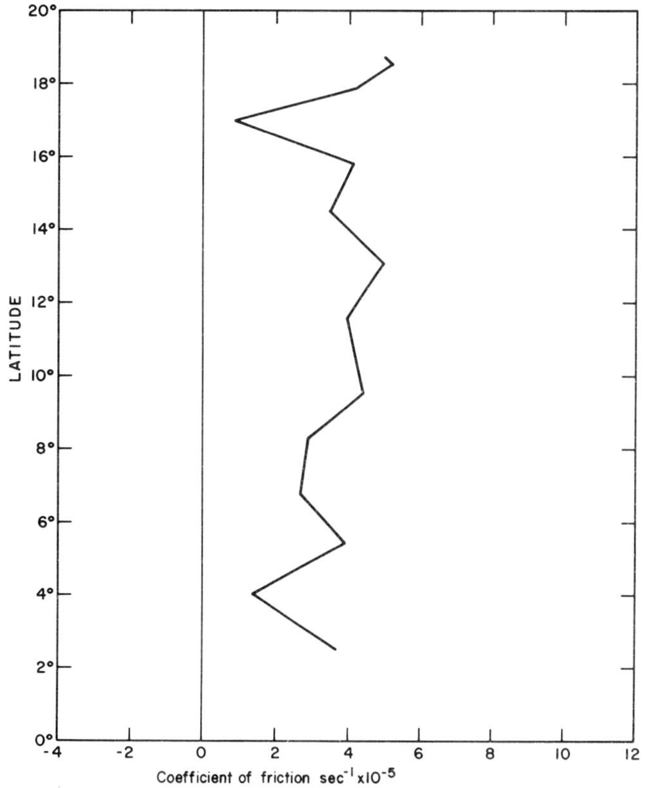

FIGURE 22. Profile of the coefficient of frictional resistance as a function of latitude averaged over 30 minutes of flying time (return flight).

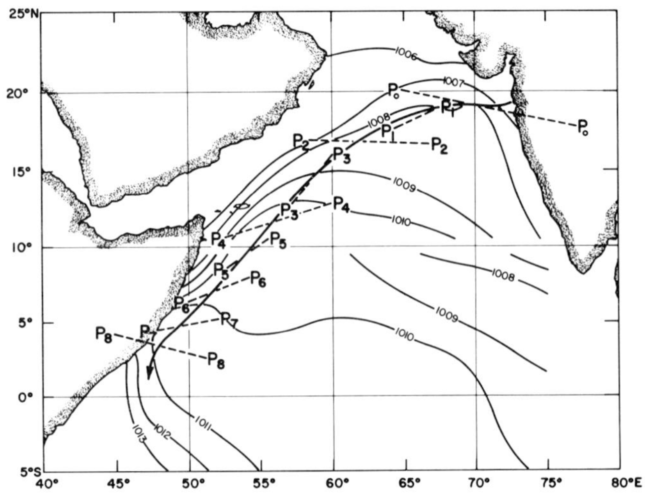

FIGURE 23. Composite pressure map during the period of the flights showing the tracks of the aircraft and the computed orientation of the isobars (outward flight).

(*h*) *Synoptic Situation during the Period of the Flights.* Figures 23 and 24 are composite pressure maps that show the tracks of the aircraft on the southward and northward flights and the approximate pressure distributions during the periods when the flights were made. The short line segments intersecting the tracks indicate the probable direction of the isobars at selected points as computed from (79) and (80). The distance separating the isobars does not represent the pressure gradient and the segments do not represent a specific pressure value. It is noted that the computed orientation of isobars along the tracks agrees reasonably well with the direction of the isobars on the analyzed charts. The period was one in which the monsoon current was developing strongly (*71*).

The great advantage of the observations of the no-drift aircraft flight lies in the fact that, assuming a steady state, the flight path reproduces the air parcel trajectories. The changing characteristics and properties of the air and the nature of the heat-exchange processes that occur along the trajectories between the air and the underlying ocean can thus be more readily studied. Aircraft and ships heading into the wind or downwind can be used to scan the pressure and wind field in the absence of a network of fixed observation stations. Such techniques may be used to considerable advantage in the future.

2. The Marshall Islands—April 1958

We noted above that with the aid of Equation (58) and a proper meteorological chart it was possible to obtain a value of our frictional coefficient. Such an exercise, we noted, leads to values of about 2 or 3×10^{-5} sec^{-1} over various regions of the oceanic tropics. In order to obtain a more accurate value we computed the speed field at 500 m over the Marshall Islands, and compared the results with the observed (mean) speed field. The relevant observations were taken from (*72*), as published *in extenso* by Joint Task Force Seven.

From the daily observations of surface pressure and upper winds, we computed the monthly (April 1958) mean surface pressure (*73*), and mean 500 m vector winds for the 10 available stations; the results are depicted in Figure 25. Next, we simulated the release of 20 parcels over a field of 5-degree squares (5–25°N, 160–175°E), and computed their subsequent motion according to the scheme already outlined above. With a preliminary value of k we compared the computed speeds with the observed ones, preferring to take our comparisons, in a particular square, from a parcel that arrived from outside the square—preferably from some distance upstream. We varied the value of k until we obtained a tolerable agreement with the observations within about 10 degrees in direction and 5 knots in speed. Three or four trials gave a value of 2.8×10^{-5} sec^{-1}; some sample trajectories are included in Figure 25, while the overall speed comparison is depicted in Figure 26. The agreement appears reasonably good; indeed, perhaps surprisingly so, for it is by no means certain that a single value of k is sufficient (*74*).

By computing the percentage change of each area originally defined by the four corner parcels of a 5-degree square, we obtained the horizontal divergence, which we analyzed as a continuous field using the successive positions of individual centroids of each square as datum points for the divergence. The results are depicted in Figure 27 along with the observed rainfall. Though we must be cautious in making such correlations, these results, in the light of other investigations (*75*), seem reasonable. Moreover, the actual divergence values are in good agreement with computations made from the winds as reported by Sinclair (*76*) for the same region.

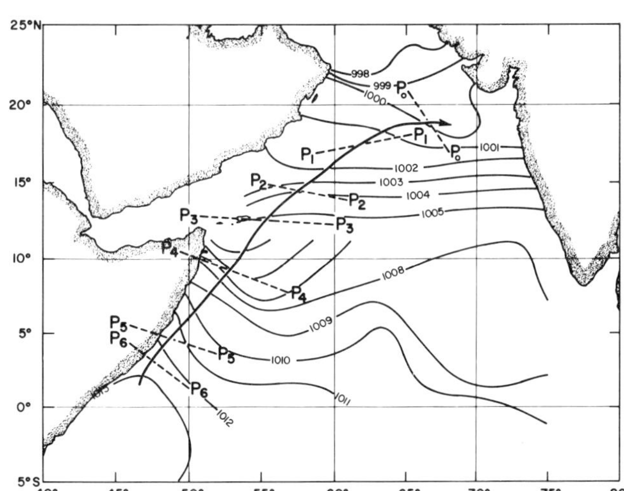

FIGURE 24. Composite pressure map during the period of the flights showing the tracks of the aircraft and the computed orientation of the isobars (return flight).

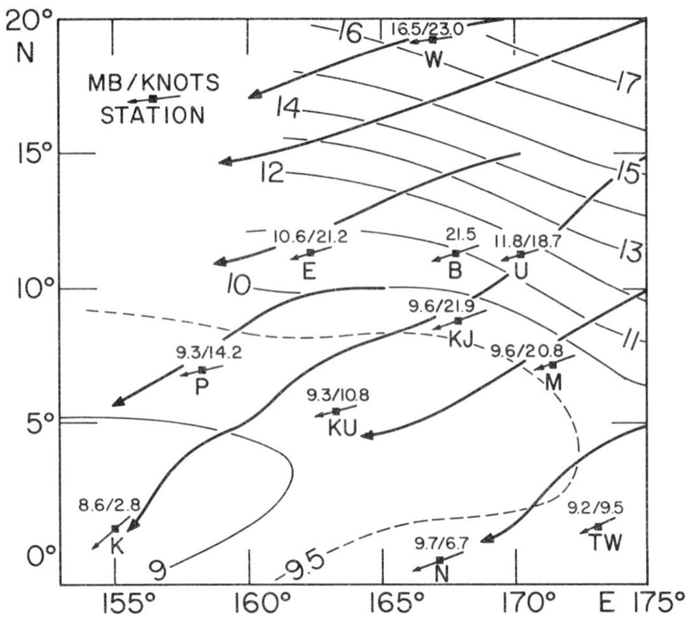

FIGURE 25. Observed mean sea-level pressure (+1,000 mb) and 500 m vector winds (knots) over the Marshall Islands, April 1958, together with computed sample trajectories. The observed pressure and mean wind speed are plotted above each station location for Wake (W), Eniwetok (E), Bikini (B), Utrik (U), Kwajalein (KJ), Majuro (M), Ponape (P), Kusaie (KU), Kapingamarangi (K), Nauru (N), and Tarawa (TW).

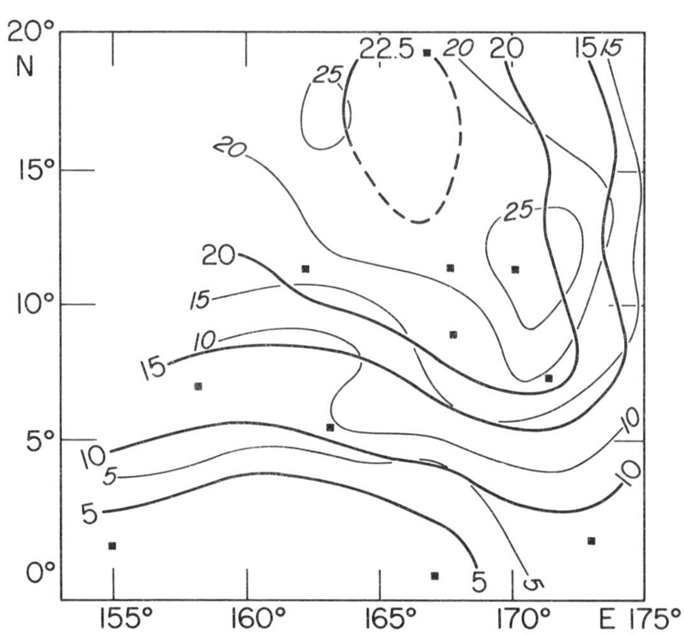

FIGURE 26. A comparison between the observed mean wind speed (heavy lines) at 500 m and computed speeds (thin lines), in knots. The speeds were computed along the trajectories of Figure 25.

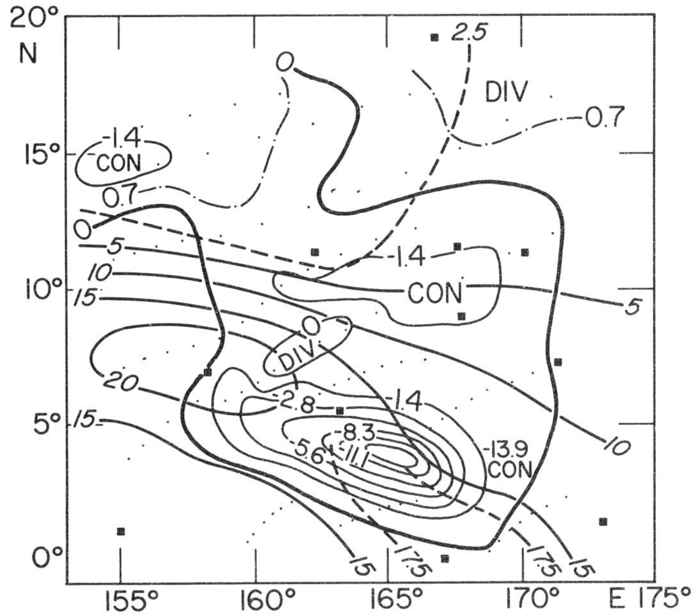

FIGURE 27. Observed monthly rainfall (inches) for April 1958 and computed horizontal divergence ($\times 10^{-6}$ sec^{-1}) determined by the displacement of areas 500 km on a side. The small dots are the locus of individual centroids of each displaced area.

We cannot, of course, claim that this particular value of k is in any way universal. Indeed, it seems plausible that k is related to the static stability, the wind speed, and the latter's variation with height, and thus is likely to vary from case to case. It is impossible to appeal to other oceanic regions (where these factors are likely to be different) for similar comparisons. We only claim that this particular value leads to reasonable comparisons in this instance, and that, for sheer lack of definitive observations elsewhere, we assumed it to apply in subsequent cases.

Though the comparison between rainfall and divergence is physically satisfying—the region of maximum rainfall had from 13 to 18 days of precipitation equal to or greater than 6.5 mm—we must retain some reservations about the region equatorward of 5°N, as noted above (*73*).

3. Indian Ocean—January

Turning to the mean conditions over the Indian Ocean in January, we relaxed our specification of the pressure field and included just the meridional variation—certainly a plausible assumption near 60°E (*77*). In addition, the available observations of (Beaufort) speed show the greatest variation of (initial) speeds to be in the meridional rather than the zonal direction. We made practical use of this assumption to adapt our original "grid" description of the Marshall Islands to a one-dimensional analogue of several parcels released from a common meridian. We may still think of moving fluid squares, but since we assumed that two parcels originally lying along the same latitude circle had the same initial speed, and were acted upon by the same pressure force, we economized and examined two parcels originally starting at two different latitudes on the same meridian. Again, our computational requirements demanded that the x-pressure gradient be the same for all y, and that the y-pressure gradient be the same for all x; though restrictive, the latter is sufficiently elastic to include several cases of general interest.

Thus, we simulated the release of seven parcels along 60°E between 15°N and 15°S, computed their individual trajectories, the horizontal divergence between mutual pairs, and analyzed the results as a continuous field. Again, the pressure gradients and initial speeds were taken from the Dutch atlas (*77*) and displayed in Figure 28, the trajectories and accompanying divergence in Figure 29. Further, for comparison, a few sample surface wind vectors taken from (*77*) have been added in Figure 29, while Figure 30 compares our computed speeds, assumed value at 500 m, and the Beaufort surface speeds of (*77*). Finally, we compared the divergence results of Figure 29 with the observed mean cloudiness, Figure 31, as deduced from averaging some forty satellite orbits during January of 1966 and 1967 (*78*).

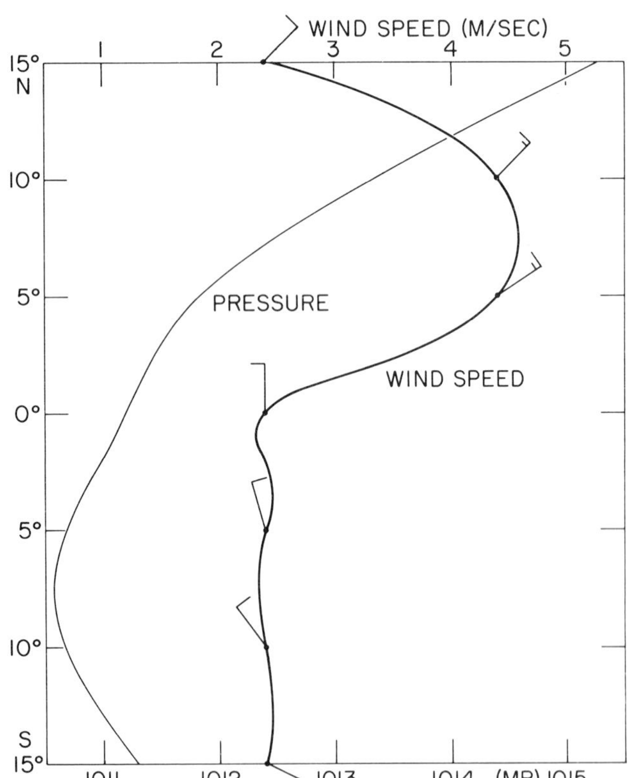

FIGURE 28. Observed meridional profile of sea-level pressure (mb) near 60E over the Indian Ocean in January, together with the profile of observed mean wind speed (m sec^{-1}). A few wind flags have been entered on the latter profile to indicate direction (north at top of diagram); each full flag represents two units in the Beaufort code. All data taken from (*77*).

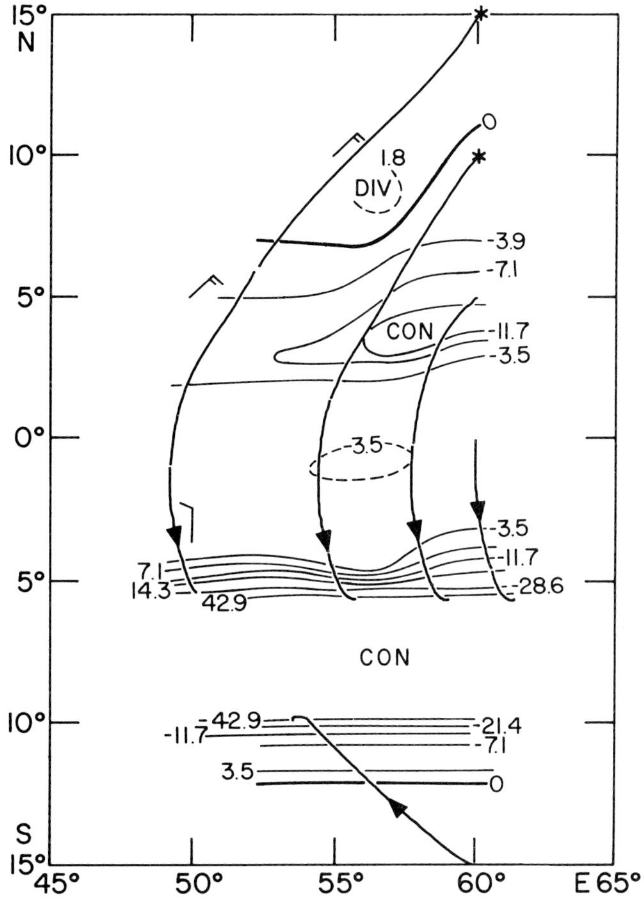

FIGURE 29. Computed trajectories and divergence ($\times 10^{-6}$ sec^{-1}), using the initial conditions of Figure 28, for the Indian Ocean, January.

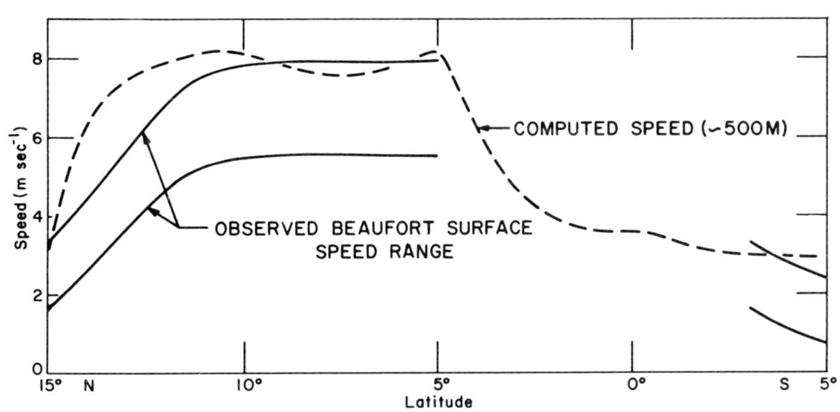

FIGURE 30. A comparison between the observed and computed speeds (m sec^{-1}) along the trajectory released at 15°N. The two solid lines represent the speed interval corresponding to the Beaufort code number along the trajectory. The region between 5°N and 3°S had insufficient data. Observed mean winds taken from (77).

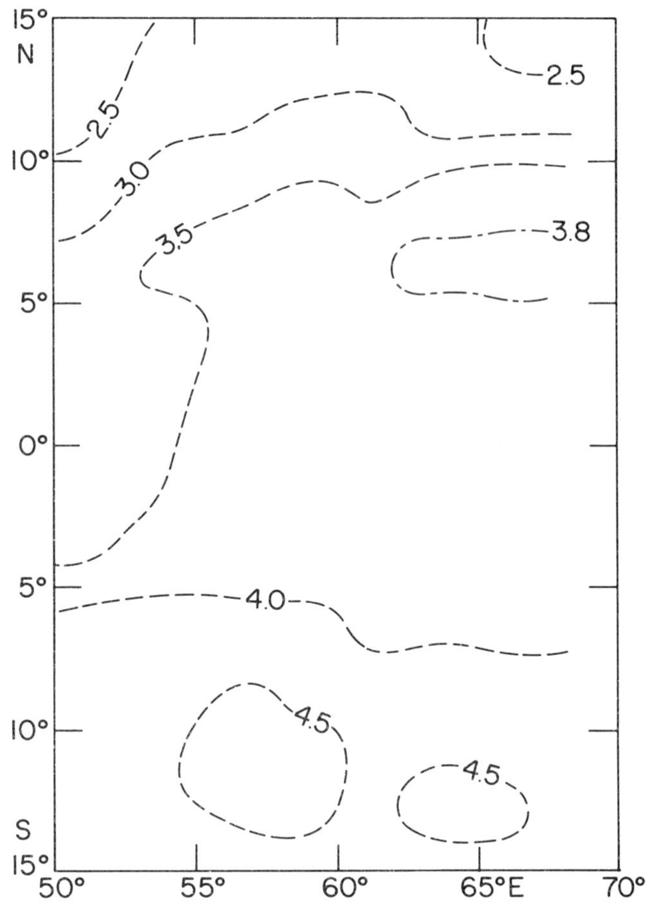

FIGURE 31. Observed satellite cloud cover (oktas) for January 1966 and 1967 (*78*).

Before discussing the results we must again acknowledge the shortcoming of comparing divergence computed from long-term climatological averages and the average cloudiness for two particular years. However, the available pressure data for January 1966 and 1967 were insufficient to provide reliable average maps, while data on mean cloudiness drawn from surface observations are too crude to provide a convincing comparison. Again, the diagnostic character of the investigation ought to be borne in mind, with some regard to the kind of detail the present technique is capable of depicting against the kind of detail we actually observe on at least some occasions. Again, we chose the highly steady regimes of the Indian Ocean, in both summer and winter, with the hope that if some interpretive success is not possible here then little is likely in less steady, or non-steady, regimes elsewhere.

Returning to the observed cloudiness of Figure 31 and the computed divergence of Figure 29, we see a broad agreement to the extent that both patterns display maximum cloudiness and convergence near 5° and 10°S with a secondary maximum of convergence near 3°N. The observed cloudiness also shows a slight maximum near 5°N. Though, the distinction to be drawn from a difference of 0.3 in the cloud cover is, understandably, open to question, the individual daily cloud cover displays this feature quite conspicuously.

4. Indian Ocean—July

Turning to the steady circulation of the summer monsoon, we again extracted mean values of the sea-level pressure and surface wind for July, choosing a meridian (90°E) where isobars trend east-west, and performed a similar series of calculations as outlined above. Figure 32 contains the observed profiles of pressure and wind, taken from the Dutch atlas (*77*).

As before, we could not compare the computed divergence—deduced from a pressure-wind map of some 50 years of observations—with similar time averages of, say, cloudiness; rather, we had to confine our comparison to the distribution of cloudiness as observed by satellite during July in both 1965 and 1966, and averaged, by 1-degree squares, by Raman (*79*). Nevertheless, Raman (personal communication) pointed out that the July surface pressure distribution during both years was very similar to the climatological average as depicted in the Dutch atlas. This might be expected from the fact that the standard deviation of surface pressure in the central portions of the Indian Ocean is about 1.5 mb.

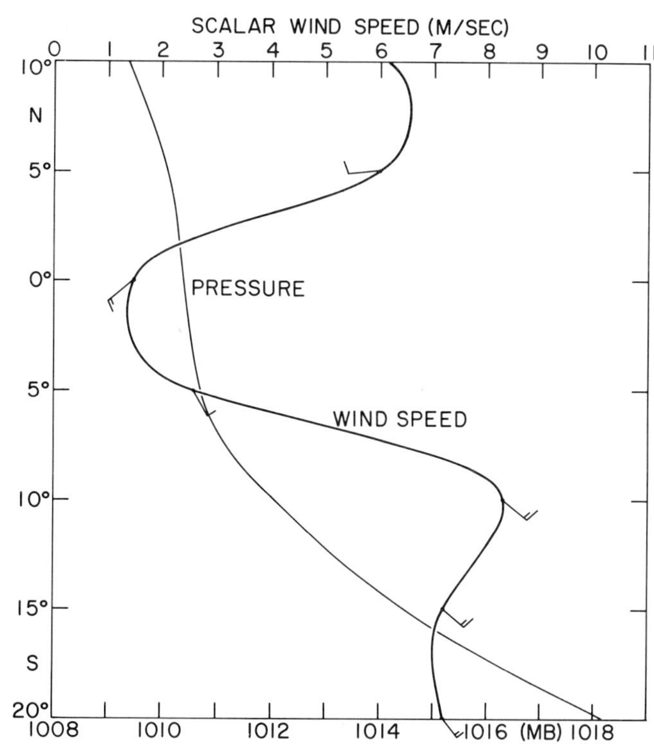

FIGURE 32. Observed mean profiles of sea-level pressure (mb) and wind speed (m sec^{-1}) near 80°E in the Indian Ocean, July. A few wind flags have been entered on the latter profile to indicate direction (north at top of diagram); each full flag represents two units in the Beaufort code. All data taken from (*77*).

Using the observed pressure distribution of Figure 32, release of air parcels was simulated along 90°E at 5-degree intervals between 20°S and 10°N, with initial wind speeds taken from the same figure. Since the Indian subcontinent intervenes near 10°N, the computations were terminated there; for clarity, only those trajectories originating south of the equator are included in Figure 33.

Again, for comparison, we included a few sample surface-wind vectors (Figure 33) from (77) together with an additional comparison of wind speed for the longest trajectory (Figure 34). The slight discrepancy between observed and computed directions near 10°N is likely the result of neglecting a very slight west to east pressure gradient in this region. Further, the computations showed that the parcels leaving 5°S and 5°N subsequently approached very closely to each other near 7°N where the agreement between their respective speeds was excellent.

In addition to the trajectories, Figure 33 contains the resulting divergence pattern. Broadly speaking, there is strong convergence just north of 10°S, a minimum just south of the equator, a narrow band of

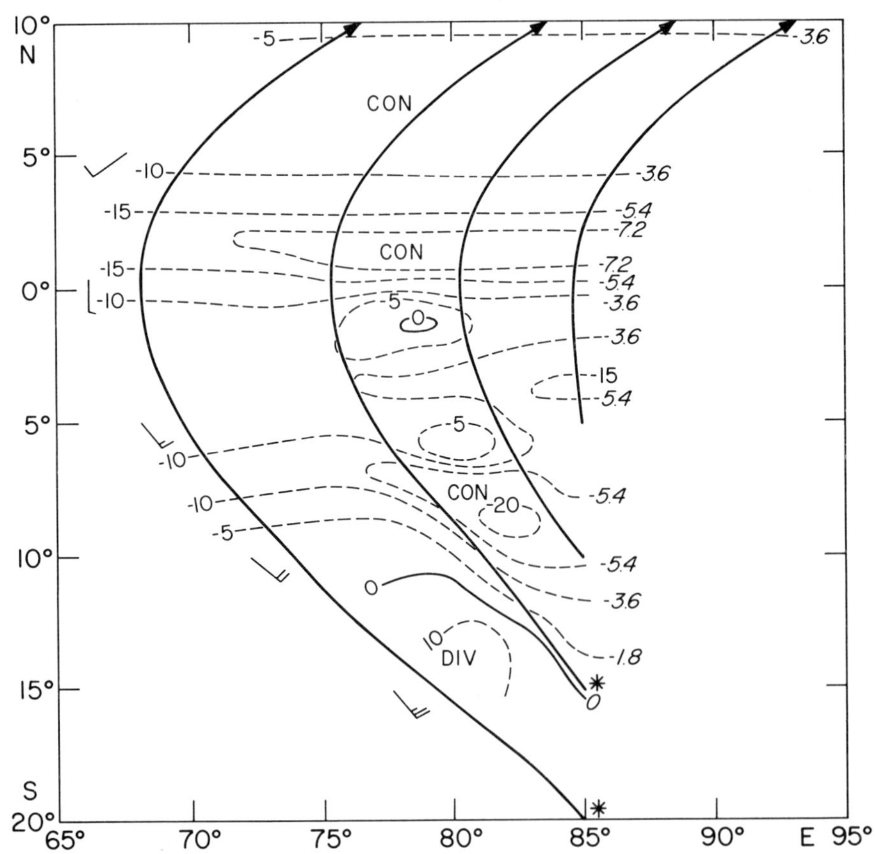

FIGURE 33. Computed trajectories and divergence ($\times 10^{-6}$ sec^{-1}), using the initial conditions of Figure 32, for the Indian Ocean, July.

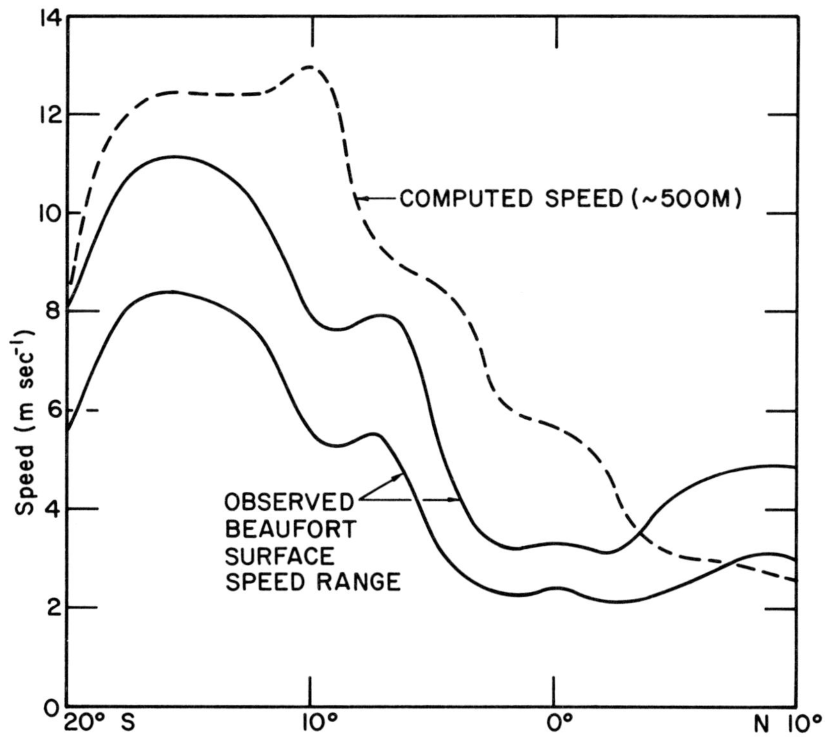

FIGURE 34. A comparison between observed and computed wind speeds (m sec^{-1}) along the trajectory released at 20°S (Figure 33). The two solid lines represent the speed interval corresponding to the Beaufort code number along the trajectory. The latter are derived from surface observations; the computed speeds are assumed to apply to a height of 500 m. Observed winds taken from (77).

strong convergence just north of the equator with values diminishing northward.

Correlation with observed cloud distribution in Figure 35 is striking. Two cloudy zones just north of 10°S and the equator are evident. Figure 35 is not just the numerical result of indiscriminate averaging, but reflects distributions observed on many individual days (*79*).

5. Gilbert-Ellice Islands—July 1964

The Gilbert-Ellice chain of islands consists of twenty-five atolls extending nearly north-south from 3°N to 10°S near 175°E; their total land area is about 325 km^2. Rainfall is measured on all atolls while on four (Tarawa, Arorae, Funafuti, Nurakita) sea-level pressure is observed (kew barometers) four times per day. In addition to these stations we may add Majuro (7°N) and Rotuma (12.5°S) giving six stations between 7°N and 12.5°S, certainly the best distribution of pressure measurements in the equatorial oceanic tropics. Lastly, Nauru and Ocean Island, on the equator, provide two pressure measurements some 5 degrees of longitude west of the major Gilbert-Ellice chain. Indeed, with these stations and others, some confidence can be attached to the surface pressure field in the region of 10°N–10°S, 170°W–160°E.

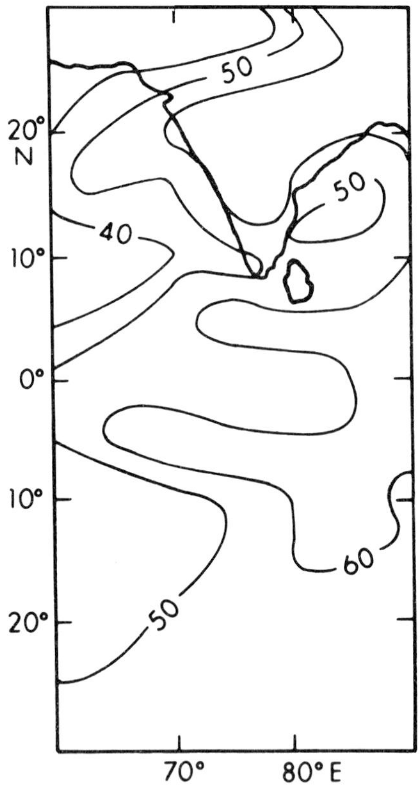

FIGURE 35. Mean cloud cover (percent) over the central Indian Ocean, as deduced from 1 × 1 degree averages of satellite observations (*79*); original data from (*78*). July 1965 and 1966.

One of the striking climatological features of the Gilbert-Ellice chain is the dramatic latitudinal variation of rainfall, a feature that has been the object of some comment for over a century. Indeed, this feature was related to scientists of the Wilkes Expedition during their visit to the Islands in 1840. It is also remarked upon in the early reports by the British Colonial Office (namely, 1910) and was subsequently documented from measurements beginning, for the most part, in the 1920s. Seelye's maps (*80*) call attention to the general features, while Figure 36 shows a profile of the observed rainfall for a particular month (July 1964), which is quite typical for other months and years.

We also undertook a latitude-time section analysis of the monthly mean sea-level pressure for the six stations (7°N–12.5°S) mentioned above. The mean values were drawn from four observations per day at all stations (*81*) for the period 1957–1965. This analysis, not reproduced here, showed that relatively high pressure on the equator usually occurs between July and October, but not every year. For the period at hand, such a distribution did occur in 1957, 1960, 1962, 1964, and 1965. Of these periods 1964 presented the best example of sustained high pressure lasting from June through October; the actual differences in sea-level pressure between the equator (Tarawa, 1°N) and 7°N (Majuro) on the one hand

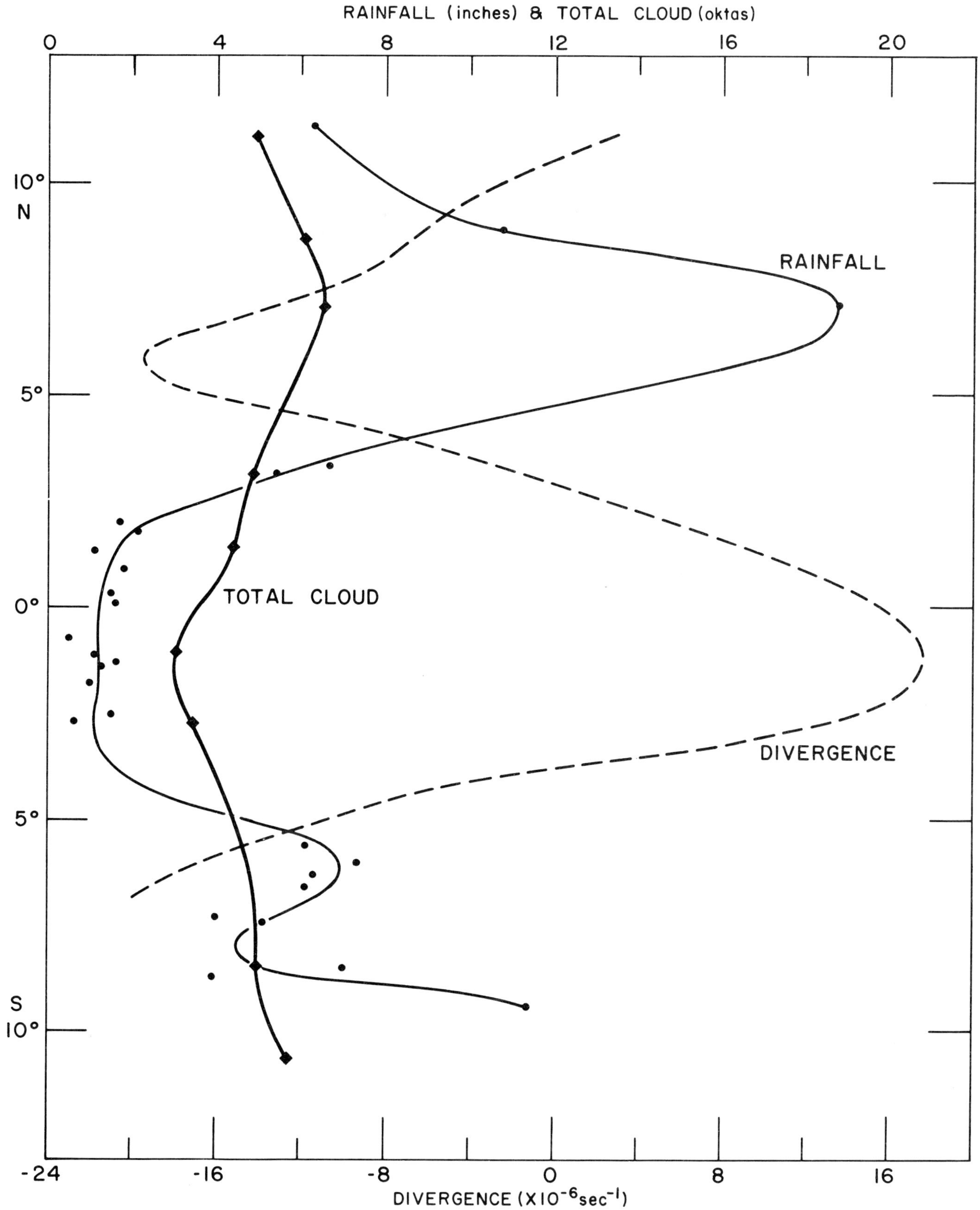

FIGURE 36. Observed rainfall (inches), July 1964, cloud cover (oktas), and computed divergence near 173°E; the latter was extracted from Figure 38. The rainfall profile is taken from 26 stations between 10°N and 9°S in the Gilbert-Ellice Islands (81) near 173°E.

and between the equator and 3°S (Arorea) on the other ranged between 0.4 and 1.0 mb. One of these 6 months (July 1964) was chosen by lot for further analysis.

The observed meridional distribution of sea-level pressure for July 1964 is presented in Figure 37; in addition, a substantial gradient along the equator of about 1 mb per 5 degrees of longitude is indicated by the values from Ocean Island. We note in passing that without Ocean Island we doubtlessly would have underestimated this zonal component.

An assessment of the initial parcel velocity was less satisfactory; we had only frequency distributions of (Beaufort) wind speed and direction

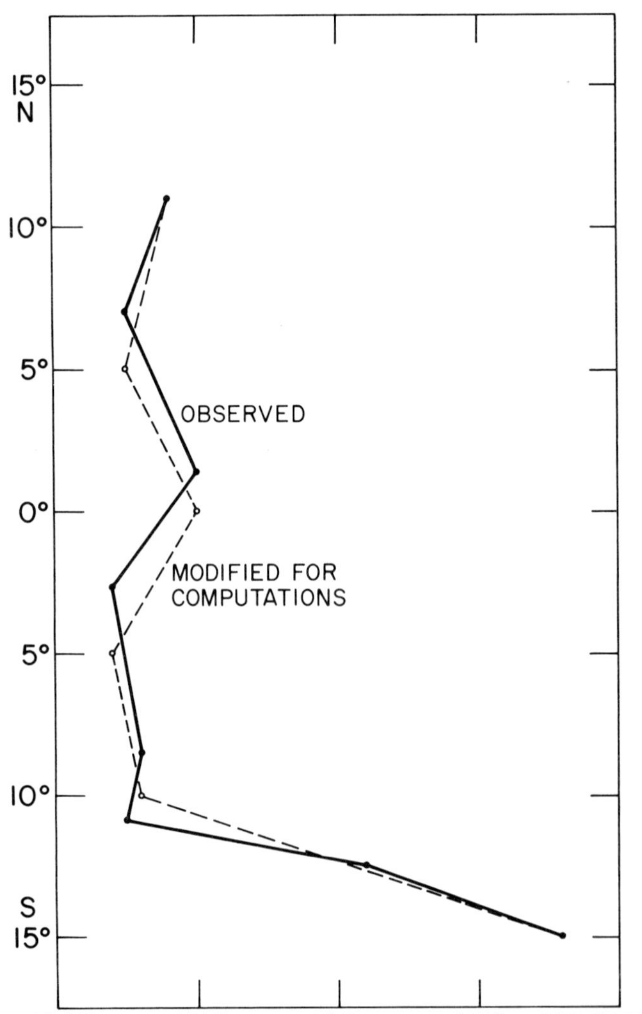

FIGURE 37. Observed meridional profile of sea-level pressure (mb) in the Gilbert-Ellice Islands for July 1964. The profile has been adjusted slightly to conform to the computational requirements of defining the gradient over 5-degree belts of latitude. In addition, a pressure gradient along the equator of 1 mb/5 degrees of longitude was indicated by the observations at Nauru and Ocean Island.

(from two observations per day) as provided in the summaries of the New Zealand Meteorological Service (*81*); these were available for the same stations observing pressure, and revealed only easterly winds during the entire month—no westerly wind being reported at any station. Since about two-thirds of the wind speeds fell in the (Beaufort) interval of 0–5.4 m sec^{-1} a uniform easterly wind of 5 m sec^{-1} was chosen for all parcels. This imprecision of the initial speeds is not likely to be a serious problem providing the results are examined after several hours have elapsed (*82*).

The actual computation simulated release of 13 parcels at 2-degree latitude intervals between 12°N and 12°S. The initial easterly wind of 5.0 m sec^{-1} ensured that all parcels started with zero divergence. The meridional pressure field, Figure 37, was adjusted slightly (the dotted line) to conform to the computational requirement of defining the pressure gradient over 5-degree latitudinal strips. Since this adjustment near 10°S could have led to a rather complicated interpretation, we did not press our observational comparisons much beyond 6°S. In addition, we assumed a zonal (east to west) gradient of 1 mb/5 degrees of latitude as suggested by the Ocean Island observations, and applied it at all latitudes, recognizing, however, that this assumption is valid over 5 to 10 degrees of longitude, beyond which we did not extend our observational comparisons.

The resulting trajectories and divergence field are presented in Figure 38. For comparison, the computed divergence along 173°E, extracted from the same figure, is displayed in Figure 36, along with observed rainfall and total cloudiness. The correlation is rather striking.

How do the computed wind speeds compare with observed values? Unfortunately, no streamline analysis is available for comparison, though observed vector winds are available (*83*) for both Majuro (7°N, 171°E) and Funafuti (8.5°S, 179°E). At 950 mb (537 m) at Majuro, observed wind of 101°, 9.5 knots, compared with the computed wind of 102°, 11.2 knots (!); at Funafuti the observed wind at 300 mb (450 m was not available) was 85°, 11.9 knots, compared with the computed value of 75°, 11.2 knots.

6. Some Theoretical Case Studies

(*a*) *Double Near-Equatorial Troughs.* Rather incomplete data have often suggested that two troughs might exist in the equatorial pressure field nearly symmetrical with the equator. Such a suggestion has sometimes been prompted by synoptic charts of the eastern Pacific, and at other times from the Gilbert-Ellice–Marshall islands. At the same time relatively higher pressure prevails on the equator (*84*). Unfortunately, the available observations are too scanty to provide sufficient data for a convincing test; nevertheless, we offer two pilot computations with double troughs symmetrically located at 10°N and 10°S with higher pressure on the equator. The gradients equatorward of the troughs are 0.5 mb per 5 degrees of latitude; the zonal gradient is assumed zero; again k is 2.8×10^{-5} sec^{-1}.

Two initial states are used: (*a*) 5 m sec^{-1} easterly winds everywhere, and (*b*) 5 m sec^{-1} westerly winds on the equatorward side of both troughs, and the same easterly winds prevailing poleward of both troughs.

The initial state, trajectories, and divergence, appear in Figures 39 and 40. The profile of horizontal divergence, taken from a common meridian to allow the motion to reach adjustment after some 20 hours, understandably is symmetrical with respect to the equator. Both initial states lead to substantially similar profiles, though divergence exists

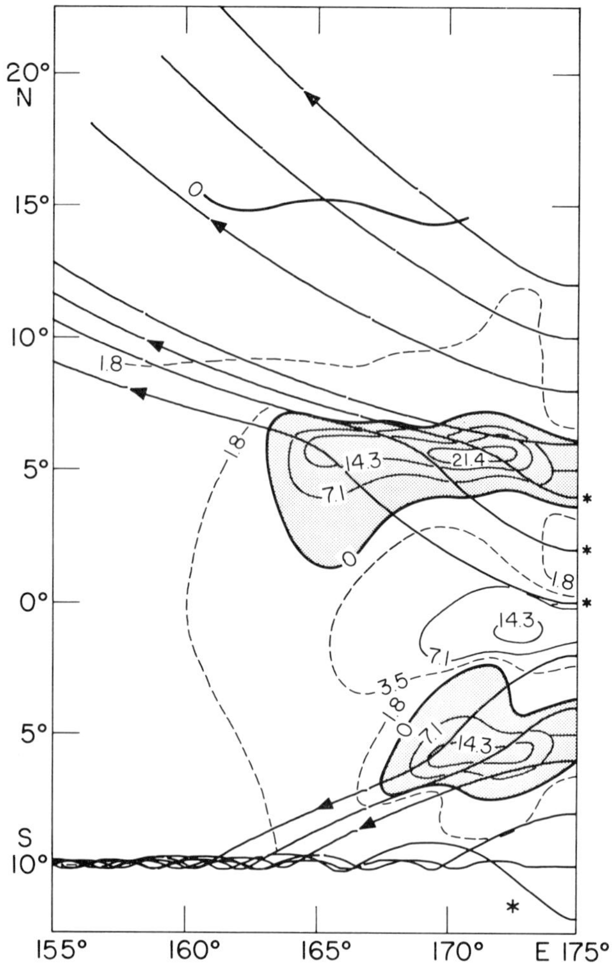

FIGURE 38. Computed trajectories and divergence ($\times 10^{-6}$ sec^{-1}) using the initial conditions of Figure 37. Initial wind speeds were assumed easterly everywhere at 5.0 m sec^{-1}. The results are assumed valid between 175° and 167°E; elsewhere the pressure gradients of Figure 37 may not apply, though the trajectories were extended over several degrees of longitude for general interest.

everywhere (8°N–8°S) in the case of easterly winds, while weak convergence zones are found near 3°N and 3°S in the case of westerly winds; further, in both cases westerly winds are found on the equatorward side of the trough, a feature Sadler and others have called attention to, though the position of the trough cannot always be determined with confidence.

A search of Sadler's cloud data (*78*) shows symmetrical cloud distribution only in the eastern Pacific during both 1966 and 1967, and only in November; unfortunately neither pressure nor wind data exist for the same period, and, besides, such symmetries could easily result from distributions of pressure and wind other than the ones considered here.

The above results do provide a useful exercise in the hazards of facile portraits of pressure and wind in low latitudes, even when friction is present.

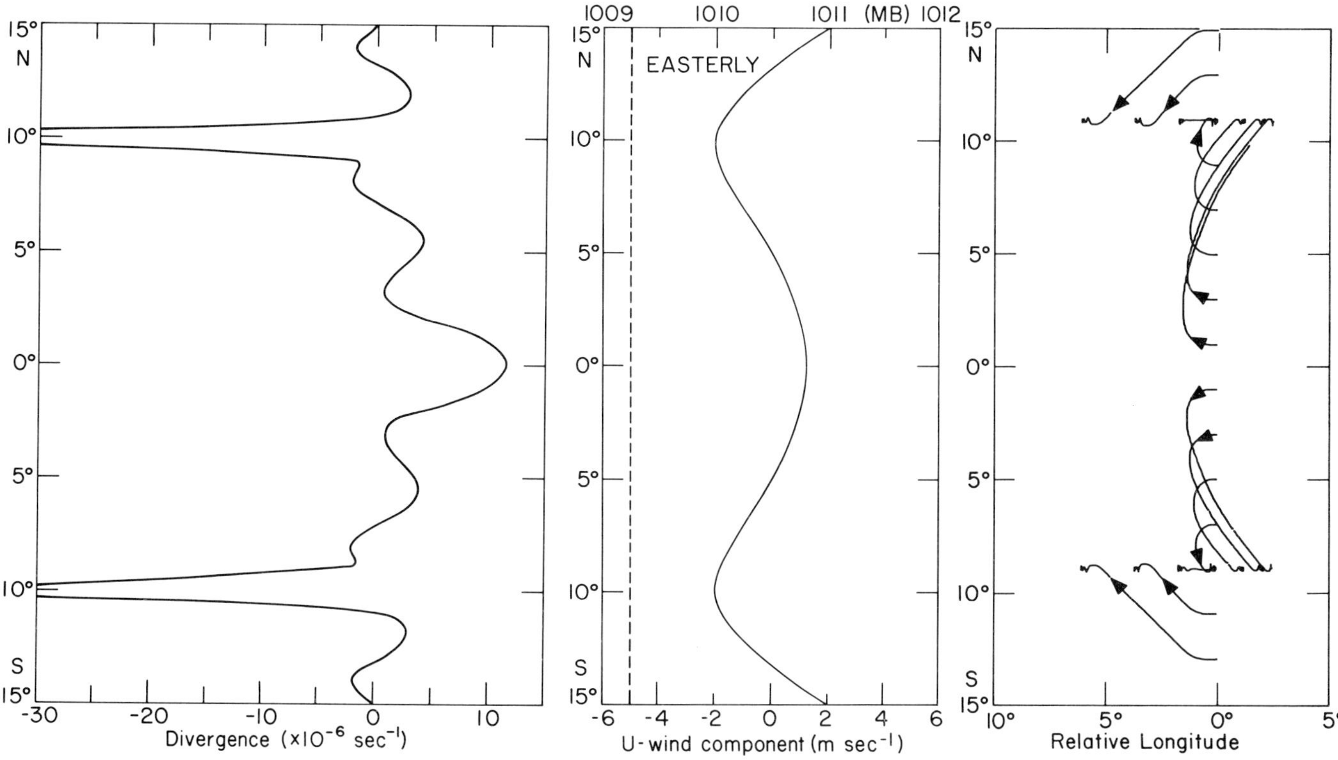

FIGURE 39. Resulting trajectories (right) and horizontal divergence (left) computed from assumed profiles of pressure and wind (center). Winds are everywhere easterly, no pressure gradient along the equator, and the frictional coefficient = 2.8×10^{-5} sec^{-1}.

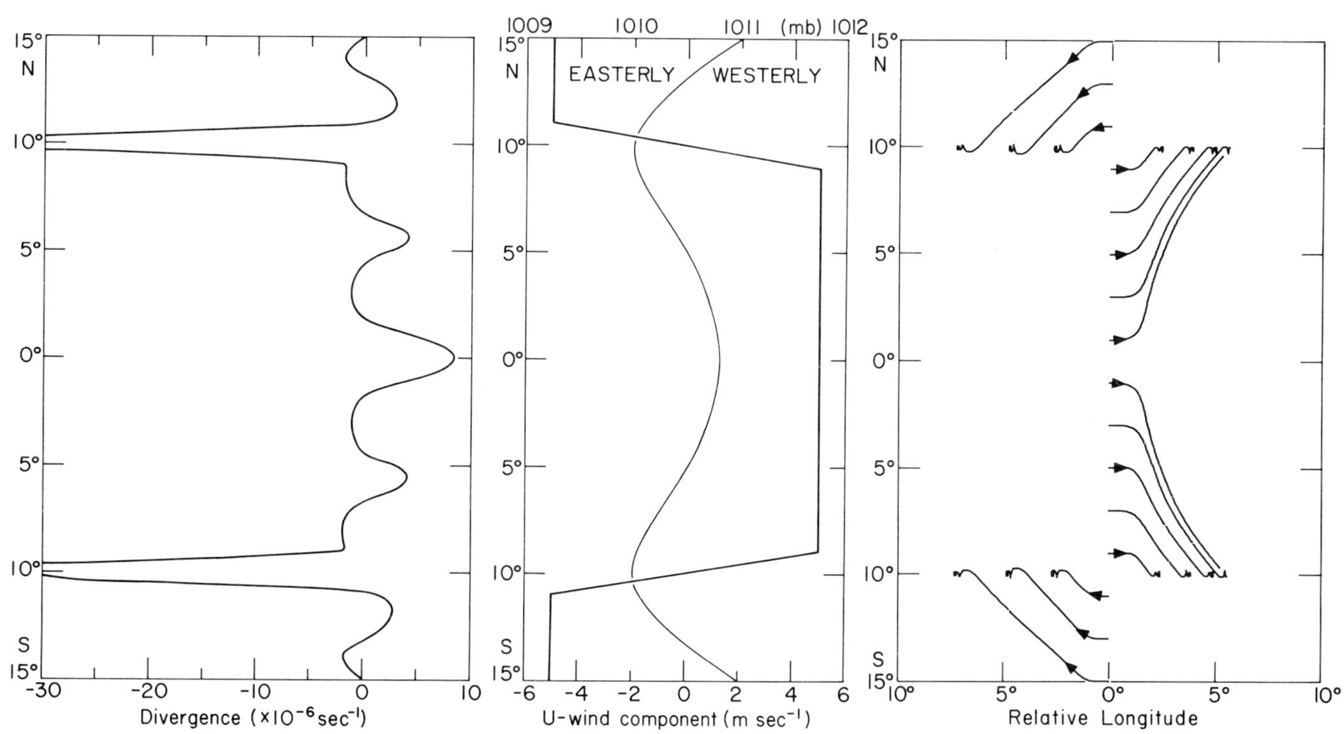

FIGURE 40. Resulting trajectories (right) and horizontal divergence (left) computed from assumed profiles of pressure and wind (center). Pressure profile is identical to that in Figure 39; the zonal winds are now westerly equatorward of 10°N and 10°S, but easterly elsewhere; frictional coefficient = 2.8×10^{-5} sec^{-1}.

(*b*) *Pressure Gradient Along the Equator.* A gradient of pressure along the equator extending throughout a wide area is not, in general, depicted in any of the marine atlases. Nevertheless, some evidence for mean zonal gradients over restricted regions (5–10° of longitude) is available from the Gilbert-Ellice Islands (quoted in section IV.B.5.), and the eastern Pacific (*85*). The latter data, summarized from three months (October-December 1955) of EASTROPIC data, show substantial zonal gradients throughout a considerable region (10° of longitude by 20° of latitude). These few observations supply some motivation for considering the following cases of zonal pressure gradients along the equator.

We considered two cases: one in which the gradient was directed

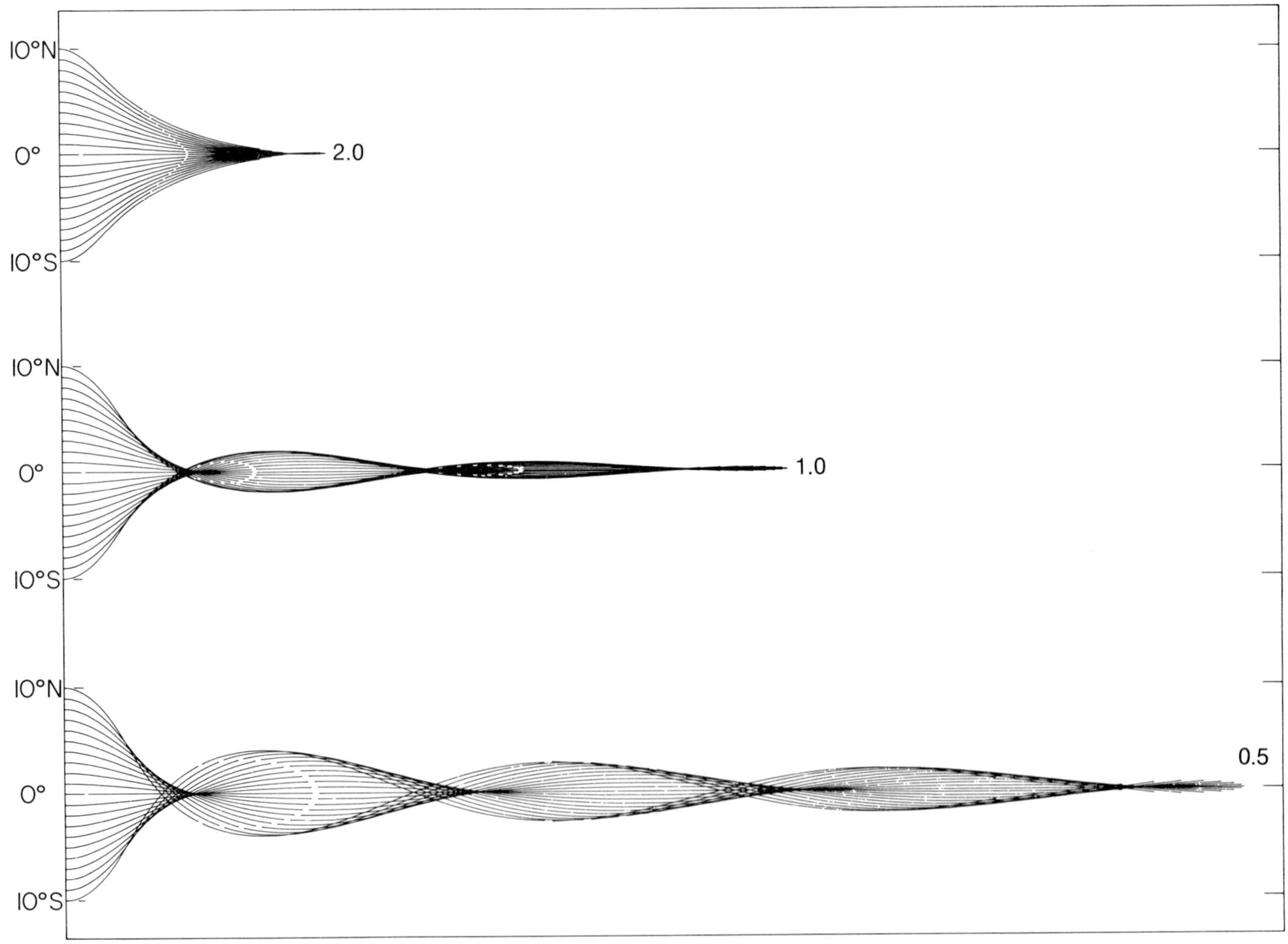

FIGURE 41. Trajectories resulting from parcels released at every degree of latitude between 10°N and 10°S with an initial speed of 5.0 m sec^{-1} in a zonal (west to east) pressure gradient of 1 mb/5 degrees of longitude. Three different values of the frictional coefficient were used, 0.5 (bottom), 1.0 (center), and 2.0 × 10^{-5} sec^{-1} (top).

along the equator from west to east (lower pressure to the east), and another opposite to this. We simulated the release of 10 parcels at 1-degree intervals 5 degrees either side of the equator using three different coefficients of friction. The initial zonal speeds were 5 m sec^{-1} and directed down the gradient in each case.

The westerly case is depicted in Figure 41, the easterly one in Figure 42. The general character of the trajectories is not difficult to understand if we recognize the relative contribution of the Coriolis acceleration acting upon parcels north and south of the equator: toward the equator in the westerly case (Figure 41), and away from it in the easterly case (Figure 42). In the former, as the particle overshoots the equator Coriolis accelerations are generated that force the parcel back toward the equator. In the latter case, Coriolis accelerations continuously deflect the parcel away from the equator. The oscillations damp with time through dissipation.

We also computed the horizontal divergence between adjacent parcels and analyzed the results as a continuous steady distribution. To avoid cluttered diagrams the analysis was omitted from Figures 41 and 42. In both cases the horizontal divergence was oriented in east-west bands: massive convergence on the equator in the westerly case, and alternating divergence-convergence bands in the easterly case (for $k = 1.0 \times 10^{-5}\ sec^{-1}$).

We do not intend to press any meteorological applications with these particular examples. Nevertheless, the general character of the westerly case is suggestive of observed water trajectories deduced from drogues in the Atlantic equatorial undercurrent (*86*); the latter, of course, apply to a very restricted region near the equator, perhaps 0.5 degrees on either side.

(*c*) *Trajectories within Some Elementary-Pressure Systems.* Algebraic expressions may be formulated in Cartesian coordinates for various elementary-pressure types such as systems constructed of isobars of circular or elliptical shape. Friction and a variable Coriolis parameter may then be incorporated by using (54) and (56).

Thus the pressure within a circular-shaped cyclone or anticyclone with equally spaced isobars may be expressed in the form

$$p = [(x - x_0)^2 + (y - y_0)^2]^{1/2} \tag{87}$$

where x_0, y_0 is the central coordinate of the circular isobaric system. Then the pressure gradient is given by

$$\frac{\partial p}{\partial x} = \frac{x - x_0}{[(x - x_0)^2 + (y - y_0)^2]^{1/2}}$$

and (88)

$$\frac{\partial p}{\partial y} = \frac{y - y_0}{[(x - x_0)^2 + (y - y_0)^2]^{1/2}}.$$

In (87) and (88), if the total pressure gradient is 1 millibar per 5 degrees of latitude,

$$P_x = \frac{-1}{\rho}\frac{\partial p}{\partial x} = \pm\frac{0.016\,(x - x_0)}{[(x - x_0)^2 + (y - y_0)^2]^{1/2}}$$

$$P_y = \frac{\pm\,0.016\,(y - y_0)}{[(x - x_0)^2 + (y - y_0)^2]^{1/2}}. \tag{89}$$

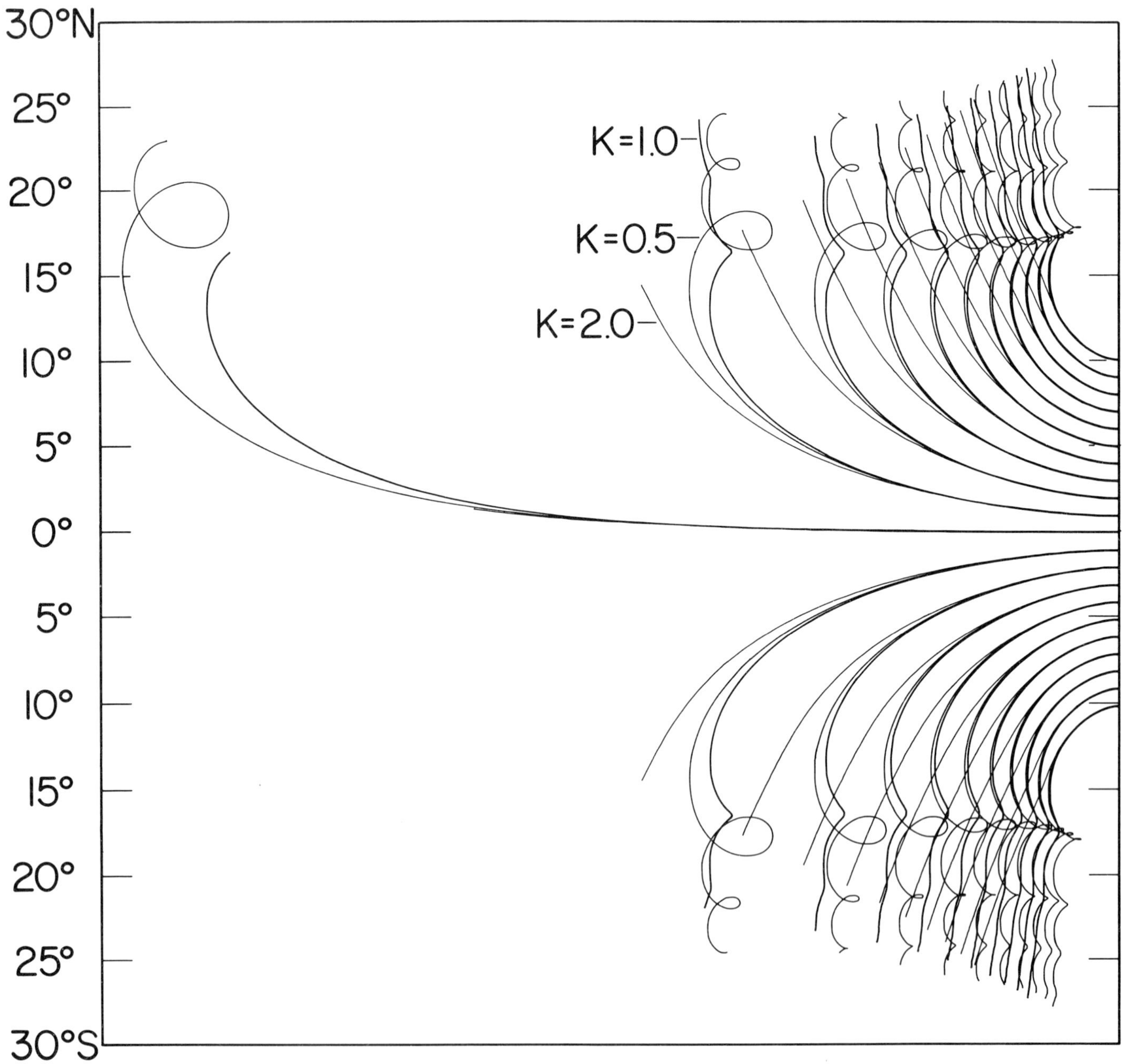

FIGURE 42. Trajectories resulting from the same initial conditions and friction as used in Figure 41, but now the zonal pressure gradient (1 mb/5 degrees longitude) has been reversed and is directed from east to west.

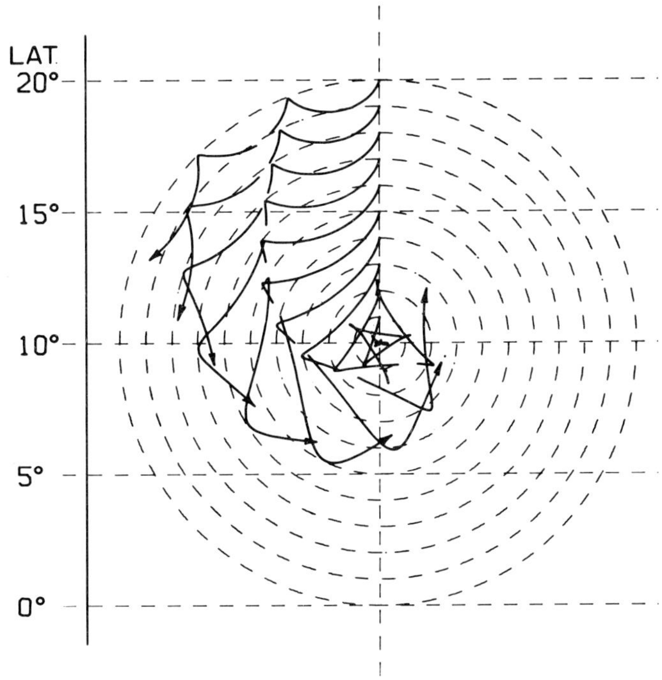

FIGURE 43. A cyclone centered at 10°N for the frictionless case when the parcels start from rest. It is noted that the oscillations increase in amplitude with decrease in latitude. The other cases of simple trajectories are not reproduced here but see Figures 48–51 for multiple trajectories around circular isobaric systems.

The constant 0.016 is a linear function of the pressure gradient in terms of the number of millibars difference in mean sea-level pressure per 5 degrees of latitude distance. (One millibar per 555 kilometers = 0.016 cm sec^{-2} approximately at normal mean sea-level pressure and temperature for the subtropics.) The positive sign in (89) would represent the anticyclonic case where pressure decreases outward from the center and the negative sign would represent the cyclonic case where pressure increases outward from the center. The expressions in (89) may be substituted for P_x and P_y in (54) and (56).

A number of cases were computed for parcels leaving at 1-degree intervals of latitude along a meridian running northward from the centers of circular-shaped anticyclonic and cyclonic systems. The systems were centered in turn at 0°, 10°N, 20°N, and 45°N. The pressure-gradient acceleration along the radius vector was always ± 0.016 cm/sec^{-2}. Initial wind speeds of 5 meters per second from the east for the cyclonic case were used; the parcels were also allowed to start from rest (Figure 43). Values of friction $k = 0$ and $k = 2.5 \times 10^{-5}$ sec^{-1} were used.

For the sake of convenience it will be useful to regard the motion as if it were in a steady state. Trajectories are then streamlines and the intersection of the streamlines, or even their confluence or diffluence, may be interpreted in the light of convergence and divergence in accordance with Equation (83).

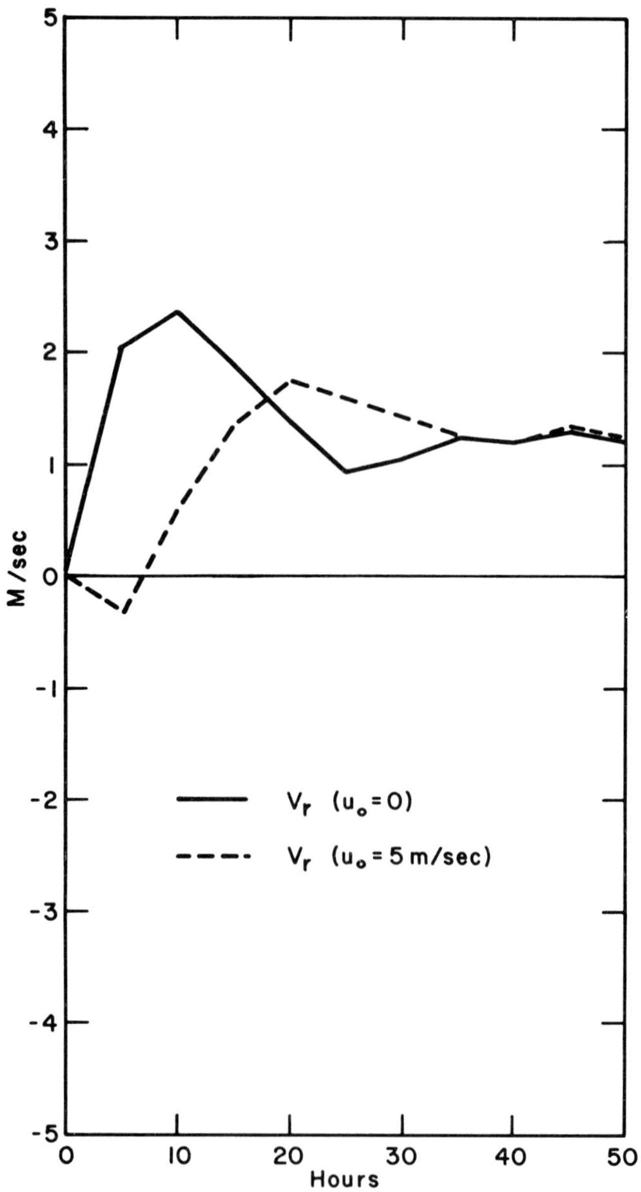

FIGURE 44. Comparison of radial velocity as a function of time of two parcels controlled by the same forces. The first parcel started with an initial wind speed of 5 m sec^{-1} and the second started from rest from 20°N with the anticyclone centered at 10°N. Frictional coefficient = 2.5×10^{-5} sec^{-1}.

In polar coordinates the wind field may be expressed by

$$\begin{aligned} V_r &= u \cos \theta + v \sin \theta \\ V_\theta &= v \cos \theta - u \sin \theta \end{aligned} \tag{90}$$

where V_r, V_θ are the radial and tangential velocities and θ is the angle between the x axis and the radius vector. Values of V_r and V_θ were computed from the u, v values that were tabulated by the computer for selected parcels taken from two cases in which friction was included. A comparison was made between the computed velocities at specified time intervals of air parcels that started with an initial wind speed of 5 m sec^{-1} on the one hand and from rest on the other. Figure 44 compares the radial velocities of two such parcels that start from 20°N within an anticyclone centered at 10°N. Figure 45 compares the tangential velocities of the same two parcels that start from 20°N on the meridian that passes through the center of the anticyclone. Figures 46 and 47 show

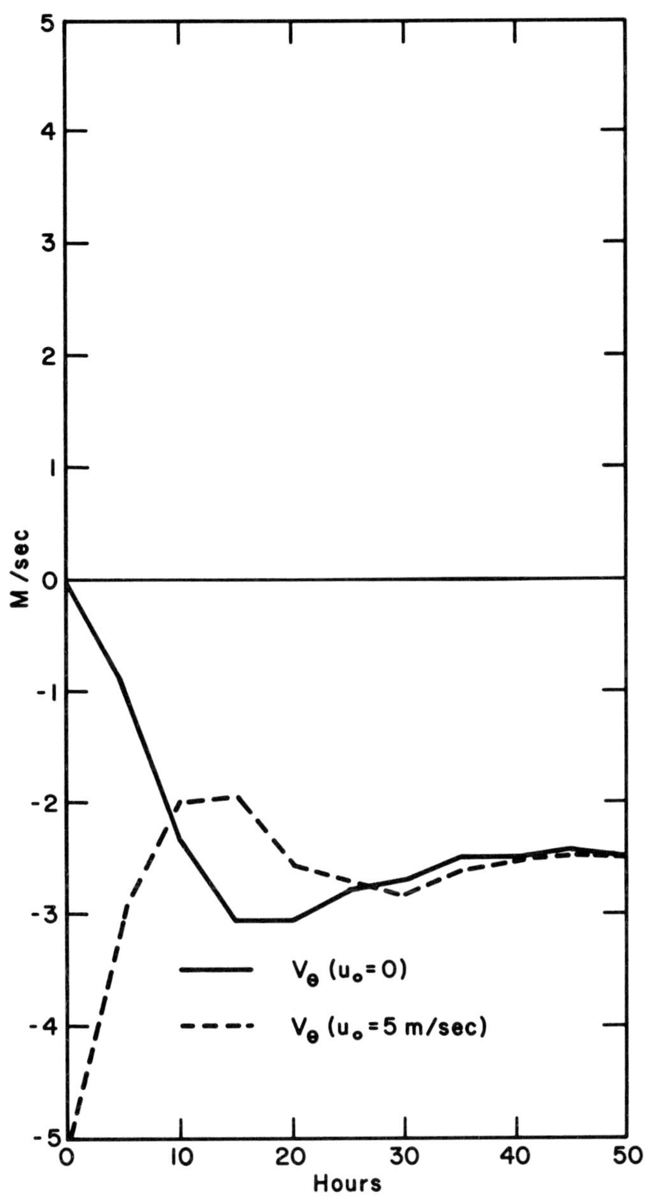

FIGURE 45. Comparison of tangential velocity as a function of time of two parcels controlled by the same forces. The first parcel started with an initial wind speed of 5 m sec^{-1} and the second started from rest from 20°N with the anticyclone centered at 10°N. Frictional coefficient = 2.5×10^{-5} sec^{-1}.

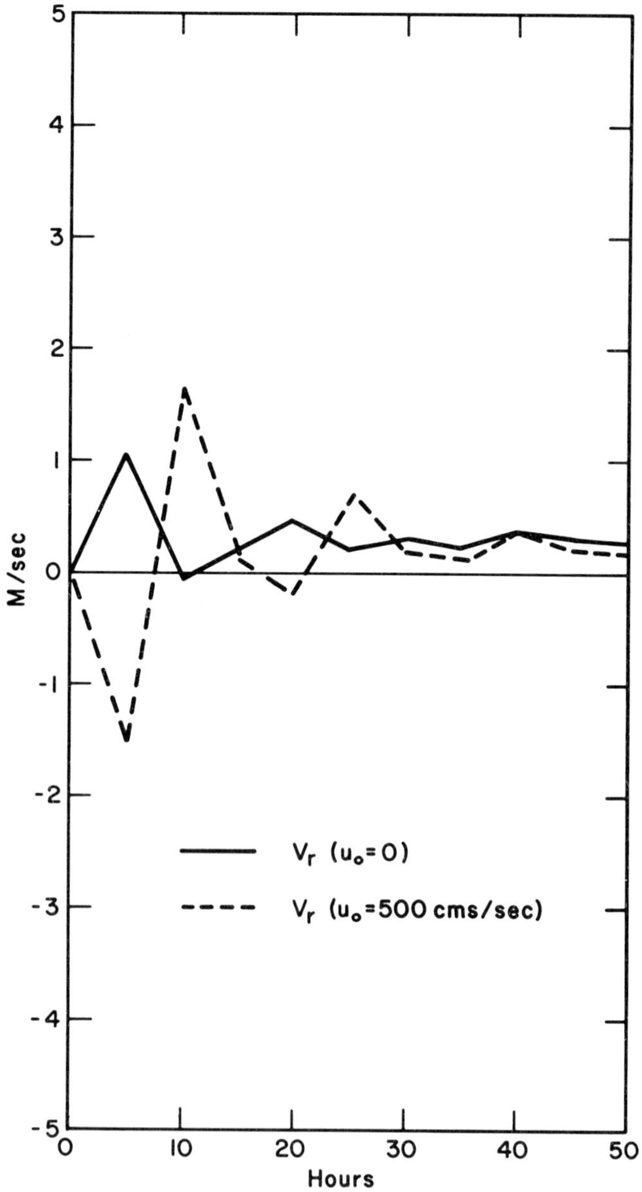

FIGURE 46. Comparison of radial velocity as a function of time of two parcels controlled by the same forces. The first parcel starts with an initial wind speed of 5 m sec^{-1} and the second started from rest from 55°N with an anticyclone centered at 45°N. Frictional coefficient = 2.5×10^{-5} sec^{-1}.

similar comparisons of the radial and tangential speeds of parcels that start from 50°N along the meridian that passes through an anticyclone centered at 45°N.

In all four cases the radial and tangential velocities and hence the wind velocities reach the same values after about 30 hours regardless of whether the initial wind speed was zero or 5 m sec^{-1}. This is perhaps a striking phenomenon that indicates the importance of friction as a stabilizing factor, constantly tending to bring wind and pressure into a balanced state. The existence of this mechanism suggests that the Lagrangian technique can predict the wind field to a reasonable degree of accuracy irrespective of possible observational or random errors in the initial conditions of either wind or pressure. The way then lies open to compute divergence and incorporate it in a numerical model of the frictional layer. The continuity equation can then be applied to obtain vertical velocities within the layer assuming that $w = 0$ at the surface of the earth.

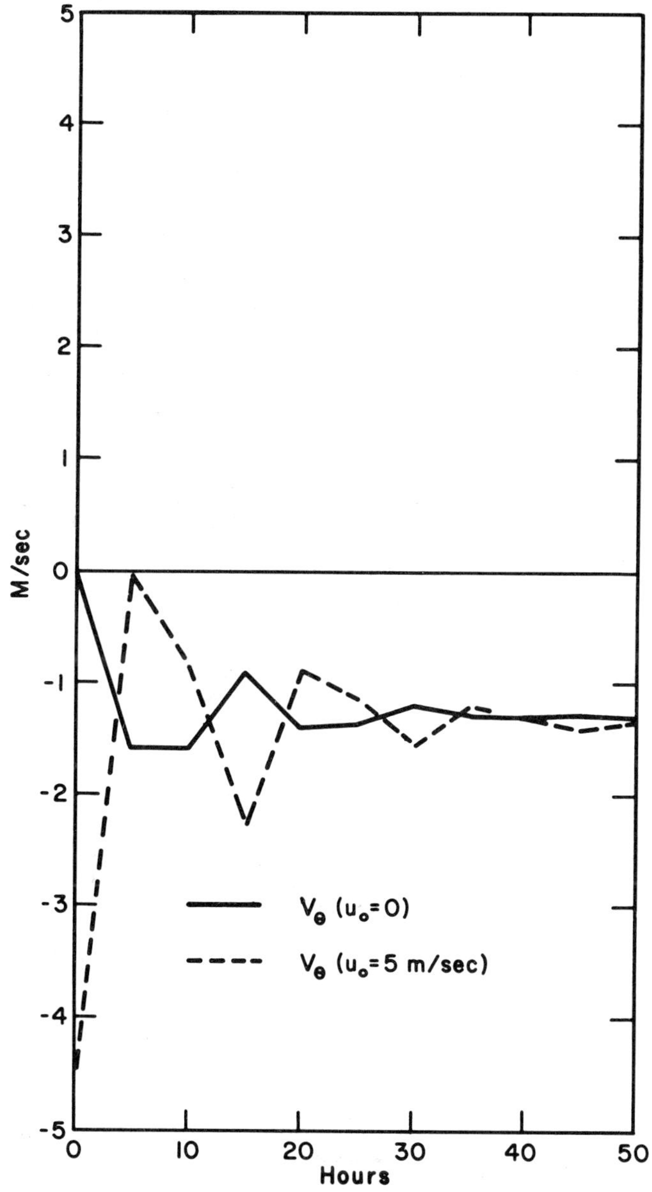

FIGURE 47. Comparison of tangential velocity as a function of time of two parcels controlled by the same forces. The first parcel started with an initial wind speed of 5 m sec^{-1} and the second started from rest from 55°N with an anticyclone centered at 45°N. Frictional coefficient = 2.5×10^{-5} sec^{-1}.

Divergence in polar coordinates is given by

$$\text{Div } \mathbf{V}_h = \frac{\partial V_r}{\partial r} + \frac{\partial V_\theta}{r\,\partial \theta} \tag{91}$$

where V_r is the radial velocity and V_θ is the tangential velocity. The first term represents the divergence along the radius vector and the second the divergence normal to the radius vector. In the cases where the isobars are circles the second term gives the divergence along the isobars. This latter term may be written

$$\frac{\partial V}{r\,\partial \theta} = \frac{\partial (r\dot{\theta})}{r\,\partial \theta} = \frac{\partial \dot{\theta}}{\partial \theta} + \frac{\dot{\theta}\,\partial r}{r\,\partial \theta}$$

$$\frac{\partial V}{r\,\partial \theta} = \frac{\partial \dot{\theta}}{\partial \theta} + \frac{V_r}{r}.$$

Thus

$$\text{Div } \mathbf{V}_h = \frac{\partial V_r}{\partial r} + \frac{V_r}{r} + \frac{\partial \dot{\theta}}{\partial \theta}. \tag{92}$$

The second term represents a field of pure divergence or convergence caused by the diffluence or confluence of the radii of the system. The third term represents the change of the angular velocity of the air parcel in a direction normal to the radius vector.

The Lagrangian form of (92) can be developed in the form

$$\text{Div } \mathbf{V}_n\, dt = \frac{d}{dr}(dr) + \frac{dr}{r} + \frac{d}{d\theta}(d\theta) \tag{93}$$

where dr and $d\theta$ are the changes in the coordinates of the parcel during the interval of time dt. We may let dt be 1 hour, as for the previous computations.

Then

$$\text{Div } \mathbf{V} = \text{Log}\,\frac{\Delta r_n}{\Delta r_{n-1}} + \text{Log}\,\frac{r_n}{r_{n-1}} + \text{Log}\,\frac{\Delta \theta_n}{\Delta \theta_{n-1}} \tag{94}$$

where

$$\frac{\Delta r_n}{\Delta r_{n-1}} = \frac{r_{i,n} - r_{(i+1),n}}{r_{i(n-1)} - r_{(i+1)\,(n-1)}}$$

and

$$\frac{\Delta \theta_n}{\Delta \theta_{n-1}} = \frac{\theta_{i,n} - \theta_{j,n}}{\theta_{i(n-1)} - \theta_{j(n-1)}}; \tag{95}$$

i, j again designate the number of the parcel that starts from some initial radial distance along radii i, j from the center of the system, and n is the number of the time step. Now if the third term in (93) is zero

$$\text{Div } \mathbf{V} = \text{Log}\left[\frac{r_{i,n} - r_{(i+1),n}}{r_{i,(n-1)} - r_{(i+1),(n-1)}}\;\frac{r_{i,n} + r_{(i+1),n}}{r_{i,(n-1)} + r_{(i+1),(n-1)}}\right] \tag{96}$$

$$\text{Div } \mathbf{V} = \text{Log}\left[\frac{r^2{}_{i,n} - r^2{}_{(i+1),n}}{r^2{}_{i,(n-1)} - r^2{}_{(i+1),(n-1)}}\right].$$

In (95) the parcel j is the parcel adjacent to parcel i on the same isobar. Parcel $i + 1$ is the parcel adjacent to the parcel i on the same radius. Equation (96) can be used if the Coriolis parameter is assumed to be constant. Then all trajectories that start from the same isobar will have the same shape. That is, the radii that pass through the two adjacent parcels will always form the same angle so that $\partial\dot{\theta}/\partial\theta = 0$.

The use of (96) enables divergence to be computed along a path defined by the midpoint between two parcels that start their motion from points along the same radius. The divergence is expressed solely as a function of the radial distances. If the term $\partial\dot{\theta}/\partial\theta$ is not zero, the computation of the divergence is more complicated and must be referred to the central point of an area formed by four parcels that enclose this area. The area in this case is distorted by the tangential velocity as well as by the radial velocity. The use of (96) does not necessitate that the Coriolis parameter be constant around the whole system but only that it be constant between two adjacent trajectories whose origins are located on the same isobar, that is, at the same radial distance from the center.

Any inaccuracy in calculating the divergence by (96) that would arise by neglect of the $\partial\dot{\theta}/\partial\theta$ term may be avoided by practical measurement and comparison of the actual areas bounded by adjacent sets of four parcels as they are traced out by the trajectories. Several exercises of this kind were undertaken. As before, the parcels were started along the meridian passing northward through the center and along other radii at 45° intervals. Figures 48–51 illustrate trajectories of air starting from rest with friction for anticyclones and cyclones centered at 10°N and 20°N. The flow for the systems centered at 10°N is directed at a greater angle across the isobars toward the center than for the systems centered at 20°N. The flow shown in Figures 50 and 51 looks remarkably like the streamlines that are drawn on synoptic charts.

The divergence field was computed for the cases shown in Figures 50 and 51 by planimetric measurement. There are a total of 40 points in each case comprising 32 individual areas for which the divergence was computed, each area being formed by two adjacent parcels along the adjacent

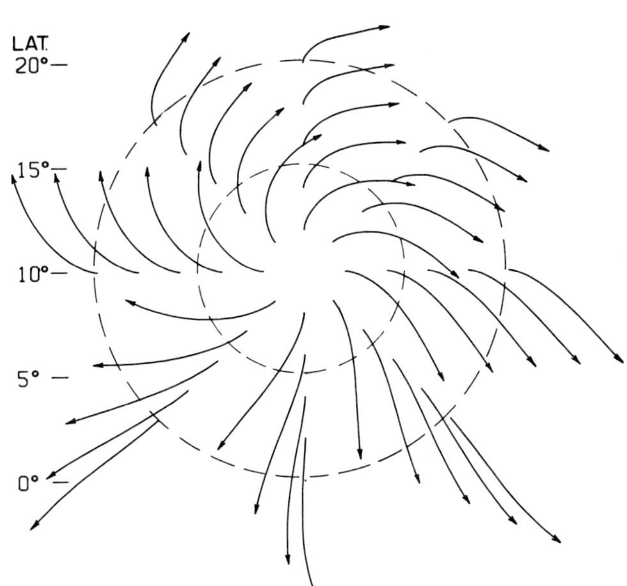

FIGURE 48. Parcel trajectories for anticyclone centered at 10°N, air starting from rest, frictional coefficient = 2.5 × 10^{-5} sec^{-1}.

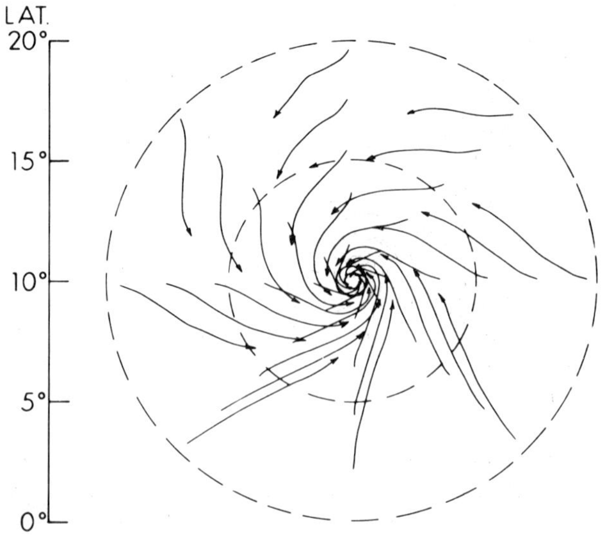

FIGURE 49. Parcel trajectories for cyclone centered at 10°N, air starting from rest, frictional coefficient = 2.5×10^{-5} sec^{-1}.

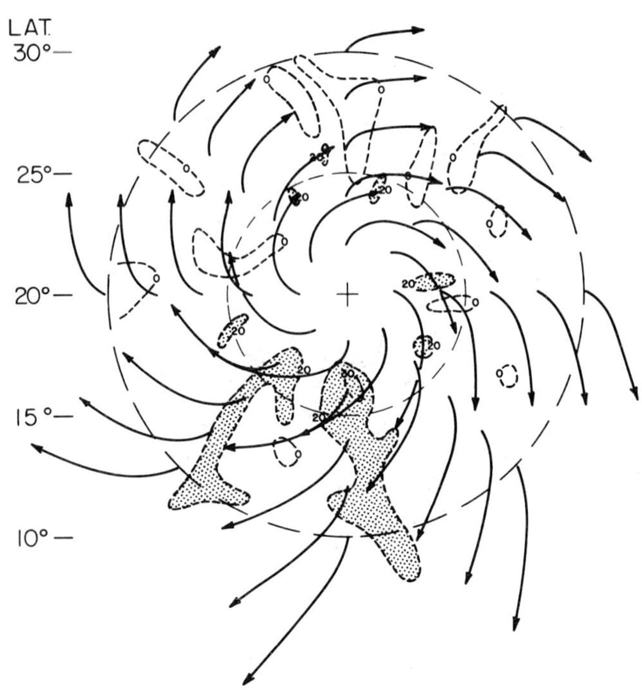

FIGURE 50. Parcel trajectories for anticyclone centered at 20°N, air starting from rest, frictional coefficient = 2.5×10^{-5} sec^{-1}. The divergence pattern computed by planimetric evaluation has been superimposed. Shaded regions indicate positive divergence. Units: $\times 10^{-6}$ sec^{-1}.

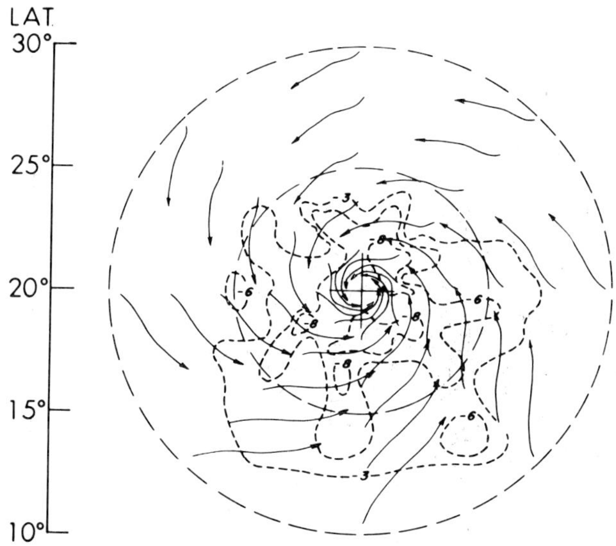

FIGURE 51. Parcel trajectories for cyclone centered at 20°N, air starting from rest, frictional coefficient = 2.5 × 10^{-5} sec^{-1}. The divergence pattern computed by planimetric evaluation has been superimposed.

radius. Each area was measured at 10-hour intervals from $t = 0$ to $t = 50$ hours inclusive. The divergence as computed from the formula Div $\mathbf{V} = \log A_2/A_1$ was then tabulated and plotted at the centroid of the area. These values were eventually plotted on an overlay of the system assuming a steady state, and the isopleths are shown on the figures. Figure 50 is for the anticyclonic case. Divergence is dominant throughout the southern sectors of the system. In the northern sectors bands of convergence and divergence alternate. In Figure 51, which refers to a low-pressure system centered at 20°N the pattern is mainly convergent, as might be expected from the general frictional inflow of the wind. In both anticyclonic and cyclonic cases the air started from rest and the value given to k was 2.5×10^{-5} $\sec^{-1}$. It is noted that the maximum convergence within the cyclonic system occurs mainly to the south of the center. A patch of convergence also appears in the northeast sector. On the northern periphery alternating bands of weak convergence and divergence are not shown. Patterns of this kind may be useful in determining preferred areas of vertical motion within anticyclonic and cyclonic systems.

Figure 52 shows the pattern of convergence for the system shown in Figure 51 from the computer printout values obtained from (96), neglecting the $\partial\dot{\theta}/\partial\theta$ term. It is seen that the patterns in the two figures are broadly similar so that it would appear that (96) gives a very useful approximation to the pattern. Equation (96) is considerably easier to handle as the values can be computed from a ratio of the squares of two lengths rather than of two areas. The evaluation of areas presented a more difficult programming task. Nevertheless, computation of the divergence field within a given layer of air would have to rely on such a program.

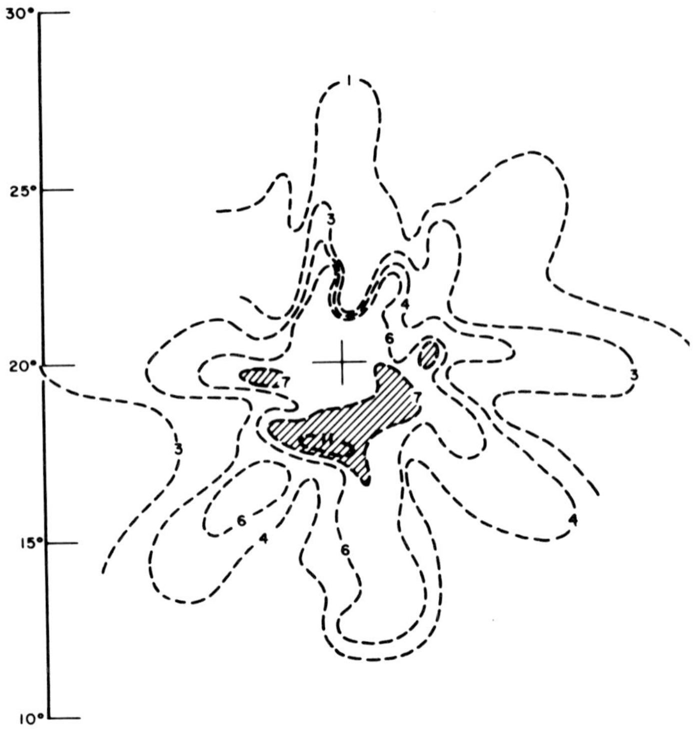

FIGURE 52. Divergence pattern for the anticyclonic case shown in Figure 50 as computed from Equation (96). Units: $\times 10^{-6}$ sec^{-1}.

We may also investigate the motion in pressure systems defined by isobars of elliptical instead of circular shape. Such systems may be expressed by

$$p^2 = \frac{(x - x_0)^2}{a^2} + \frac{(y - y_0)^2}{b^2} \tag{97}$$

where a and b are constants defining the eccentricity of the ellipse. The pressure-gradient field is then given by

$$\begin{aligned} \frac{\partial p}{\partial x} &= \pm \frac{x - x_0}{\left[\dfrac{(x - x_0)^2}{a^2} + \dfrac{(y - y_0)^2}{b^2}\right]^{1/2}} \\ \frac{\partial p}{\partial y} &= \pm \frac{(y - y_0)}{\left[\dfrac{(x - x_0)^2}{a^2} + \dfrac{(y - y_0)^2}{b^2}\right]^{1/2}}, \end{aligned} \tag{98}$$

P_x and P_y containing the constant 0.016, or some multiple or fraction of this may then be substituted in (54) and (56) as before. The ellipse is a useful case to study since subtropical anticyclones are frequently of such a shape. Thus, if, for example, we set $a = 4$, $b = 1$ the resultant pattern for positive values of P_x and P_y will resemble an elongated subtropical anticyclonic cell. If $a = 1$, $b = 4$ the pattern will resemble an elongated north-south-oriented ridge.

Figure 53 shows parcel trajectories for the latter type of pressure distribution. The parcels start from coordinates spaced at 1-degree intervals of latitude along the meridian that runs northward from the center of the system. At 10°N, $k = 2.5 \times 10^{-5}$ sec^{-1} and the air starts from rest. There is confluence of streamlines. Figure 54 shows trajectories in an elongated subtropical type of anticyclone created by reversing the magnitudes of a and b in Equation (98). The parcels start from rest at 1-degree intervals of longitude along the 10-degree parallel of latitude. In the latter two figures the isobar spacing is 1 mb/5° latitude along the major axis.

The divergence pattern computed by planimetric measurement of areas in the manner described for the two circular anticyclonic cases has been superimposed on Figure 54. The pattern is similar to that exhibited by the circular-shaped anticyclones with divergence occurring throughout most of the system, apart from a region of weak convergence in the northerly sector. This pattern resembles that which might be expected from actual average-weather conditions experienced within subtropical anticyclonic systems.

In fact Ramage (*87*) finds that the subtropical ridge in the southern Indian Ocean coincides with the axis of maximum wind divergence. The latter axis lies about 1,000 km south of the axis of minimum rainfall

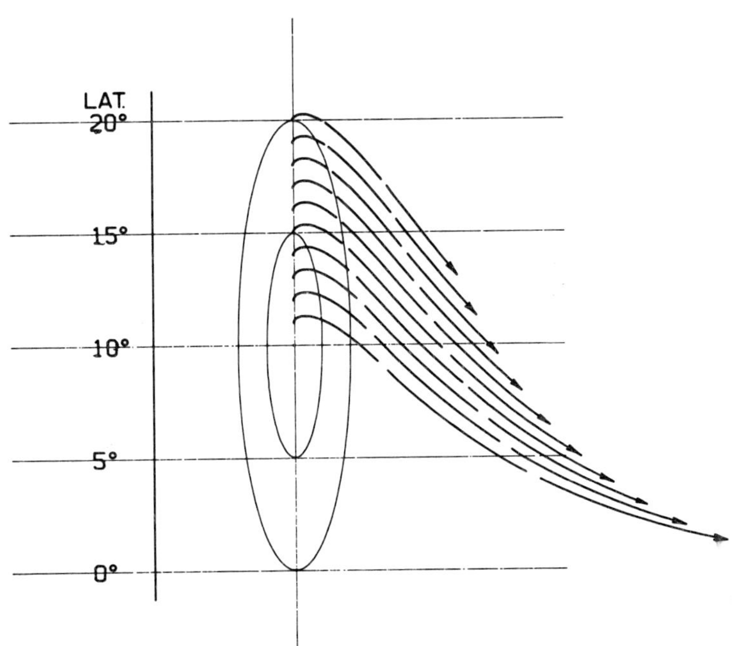

FIGURE 53. Particle trajectories for the case of an anticyclone with its ridge oriented north-south, centered at 10°N, frictional coefficient = 2.5 × 10^{-5} sec^{-1}, air starting from rest. Isobaric spacing is 1 millibar.

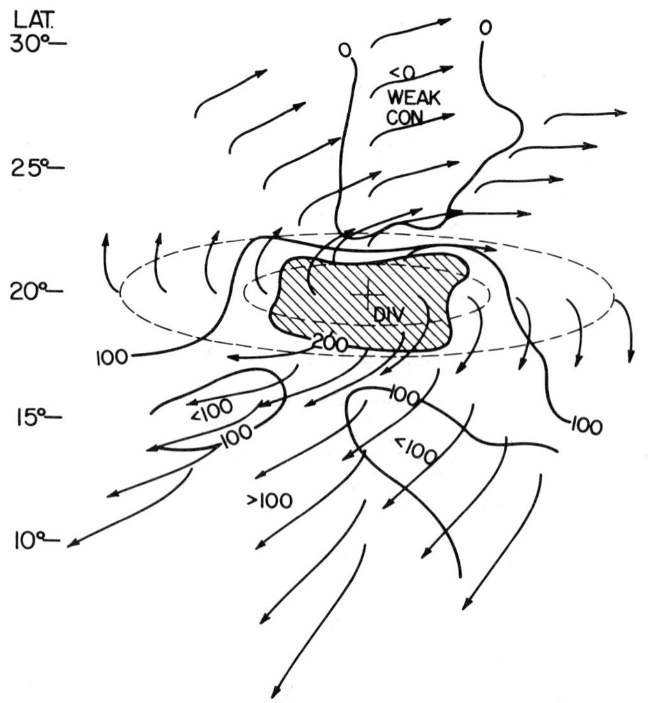

FIGURE 54. Particle trajectories for the case of an anticyclone with its ridge oriented east-west, centered at 20°N, frictional coefficient = 2.5×10^{-5} sec^{-1}, with the divergence pattern superimposed. Isobaric spacing is 1 millibar. Labeled isopleths times $.25 \times 10^{-6}$ gives horizontal divergence in units of 10^{-6} sec^{-1}.

frequency. He suggests that the divergence, acting for some time on air outflowing from the ridge, produces a downstream lowering of the inversion and precipitation decrease.

7. Numerical Steady-State Frictional-Layer Trajectories and Weather

An adequate working numerical model of the frictional layer of the atmosphere must be formulated over a grid-point network. The computations of the trajectories described in the preceding sections of this monograph have been restricted to special cases where the pressure gradient varied with latitude but not with longitude. In Section 6 the latter restriction was removed when cases were considered where the pressure field could be expressed by equations for a circle or an ellipse. For the general case the pressure field should not be restricted in any way. A grid of points should be used for the input of the initial conditions for both pressure gradient and wind field given by the u, v components of the velocity of the wind. Such a field of points will then move in accordance with Equations (54) and (56) with appropriate values of the coefficient of frictional resistance. The first simple experiment with a grid of points was described in section IV.B.2. The encouraging results obtained with this grid in the Marshall Islands area served as a starting point for further experiments over a much wider geographical area of the synoptic chart. First trials allowed the air parcels that were released from the corners of 5-degree squares of latitude and longitude to run for a period of time, say 50 or 100 hours, after which a count was made of the density of parcels within grid squares. The latter method is somewhat crude but reasonable results were obtained for several cases over the Pacific and Indian oceans regions. Areas where the density of parcels was high corresponded with areas of above average cloudiness and precipitation. Conversely, areas with few parcels were associated with little cloudiness and precipitation.

An experiment embodying the computer evaluation of the changing areas bounded by adjacent groups of parcels originally located at the corners of the 5-degree squares was performed over the Indian Ocean using the mean pressure and wind values from the Indian Ocean Climatological Atlas (*88*) for July 1963 (Figure 55). In this case the initial conditions incorporated the mean u, v wind-speed components for selected grid points. Figure 56 shows the computed divergence pattern while Figure 57 shows the observed pattern of the frequency of precipitation as obtained from an analysis of ships' observations. The main features of the divergence pattern are a belt of divergence extending from the East African coastline east northeastward. This belt reaches a maximum at about 10°N, 60°E. Northeastward of this locality divergence decreases in magnitude and reverses sign north of 15°N and east of 65°E. A center of convergence is located just west of Bombay. A region of convergence covers the area of the equator between 50° and 70°E. South of about 12°N and east of 65°E the pattern is weakly divergent. The convergent area off the South Arabian coast is probably an effect created partly at least by boundary conditions. The rainfall pattern of the frequency of ships'

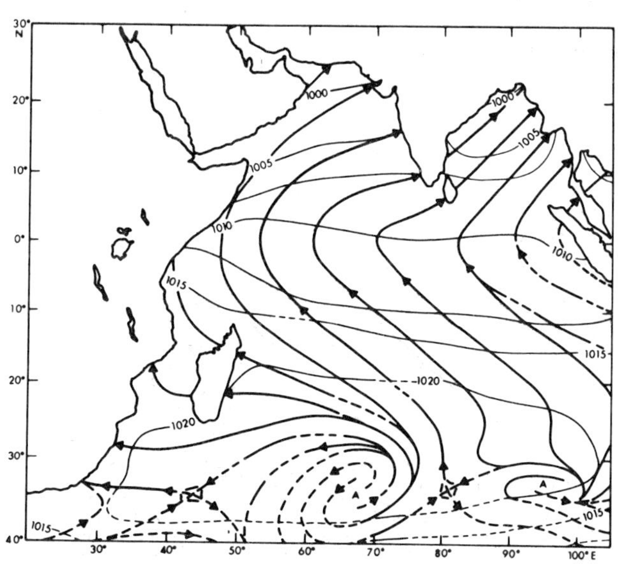

FIGURE 55. Mean surface pressure distribution and streamlines for the Indian Ocean for July 1963.

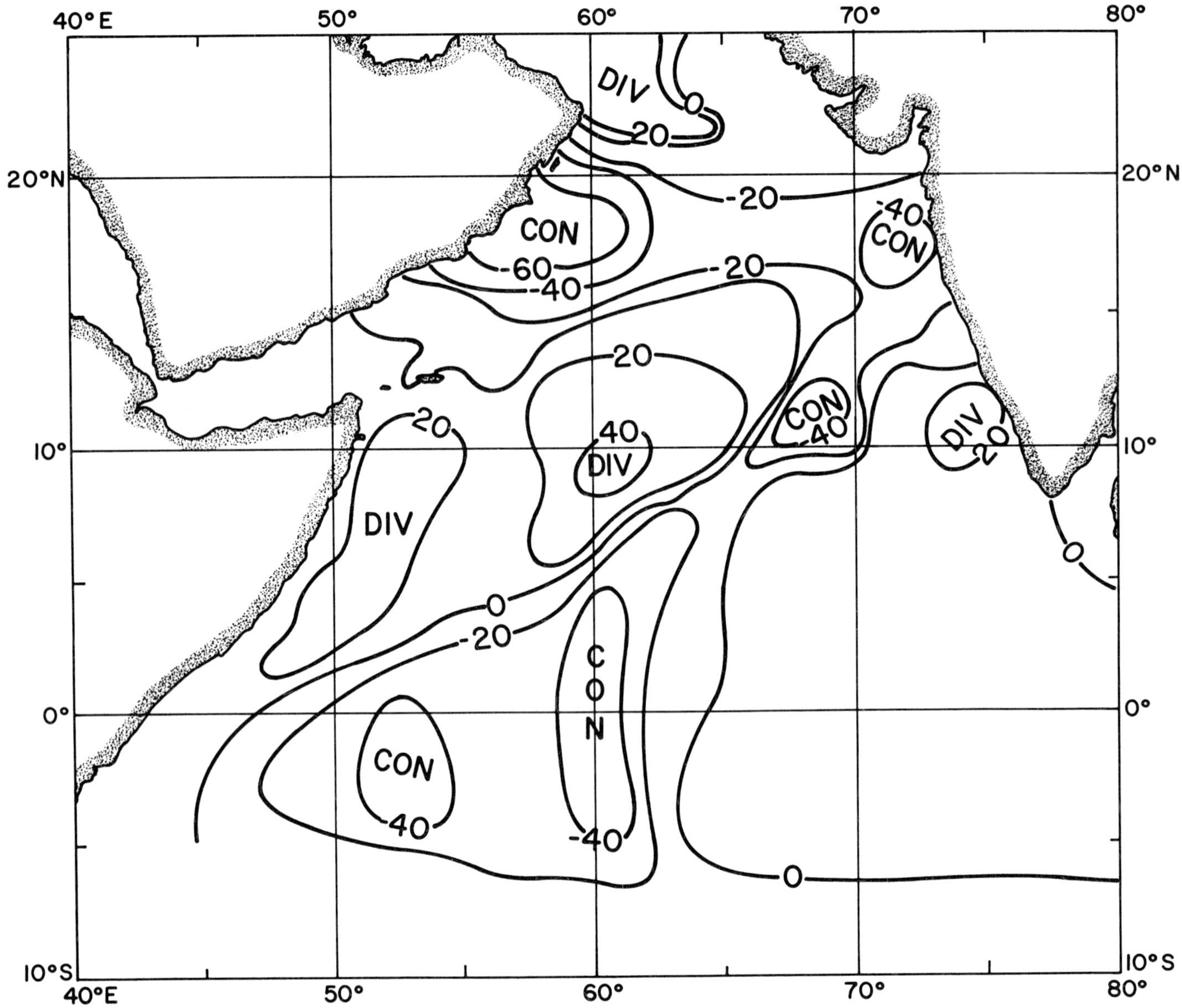

FIGURE 56. Computed divergence pattern for the Indian Ocean for July 1963. Units: 10^{-6} sec^{-1}.

observations reporting precipitation agrees in many respects with the computed divergence pattern. The area of maximum precipitation frequency extending south southwestward from Bombay is well correlated with the convergence belt. The dry area over the western part of the Arabian Sea falls broadly in a divergent zone. Elsewhere, except near boundaries, there are no wide discrepancies between the divergence field and the precipitation pattern.

Secondly an actual synoptic case was run. The composite pressure distribution obtained from the synoptic map shown in Figure 23 was used to obtain grid-point values of the pressure distribution. The initial wind was assumed to be calm in this case. Figure 58 shows the computed divergence pattern. Broadly speaking it is fairly similar to the mean pattern computed for July 1963 shown in Figure 56. Observations of weather for the subsequent hours indicated cloud and rainfall at Bombay and along the Indian west coast. There was little evidence of precipitation over the Arabian Sea west of 70°E over the computed divergent areas.

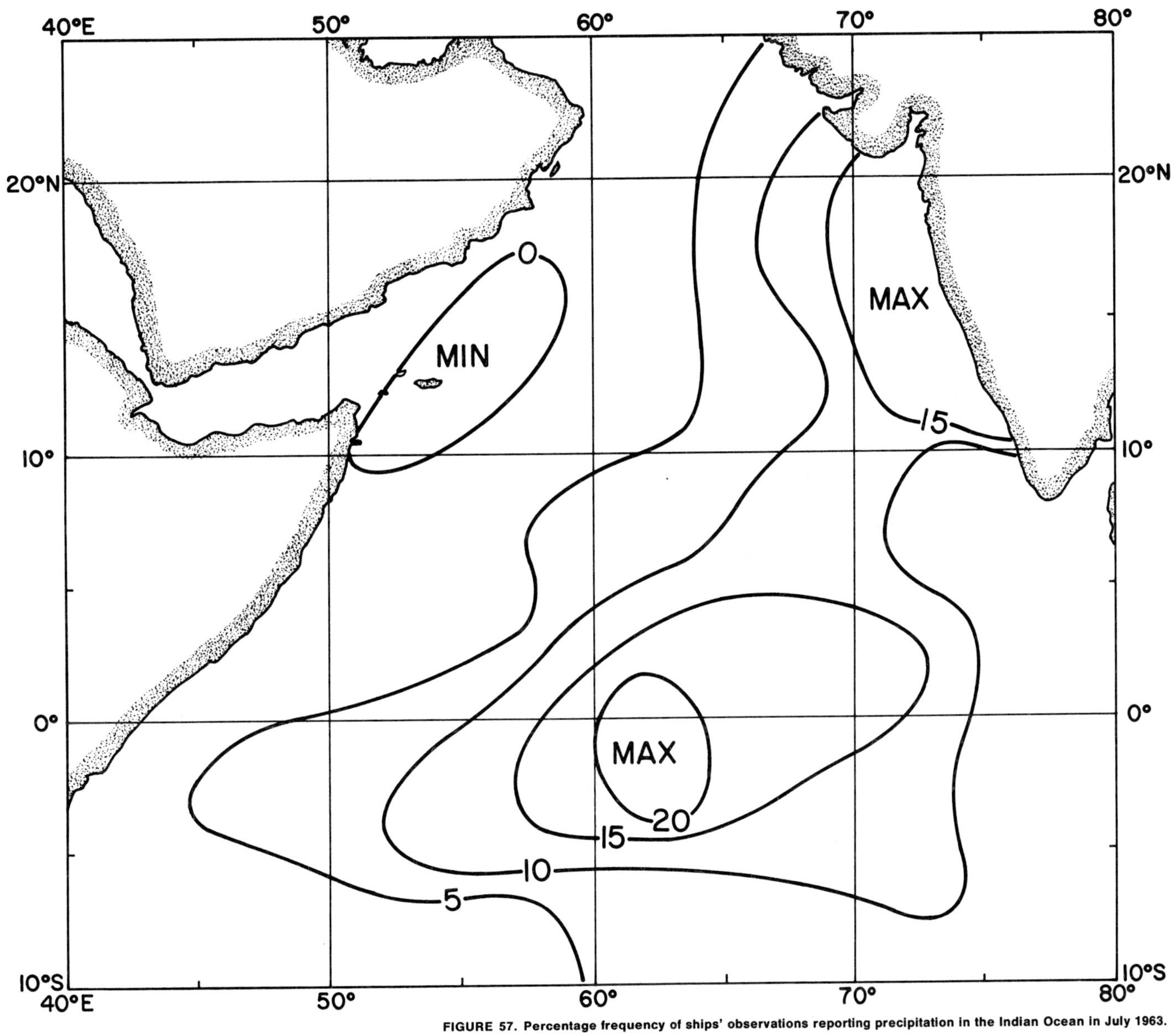

FIGURE 57. Percentage frequency of ships' observations reporting precipitation in the Indian Ocean in July 1963.

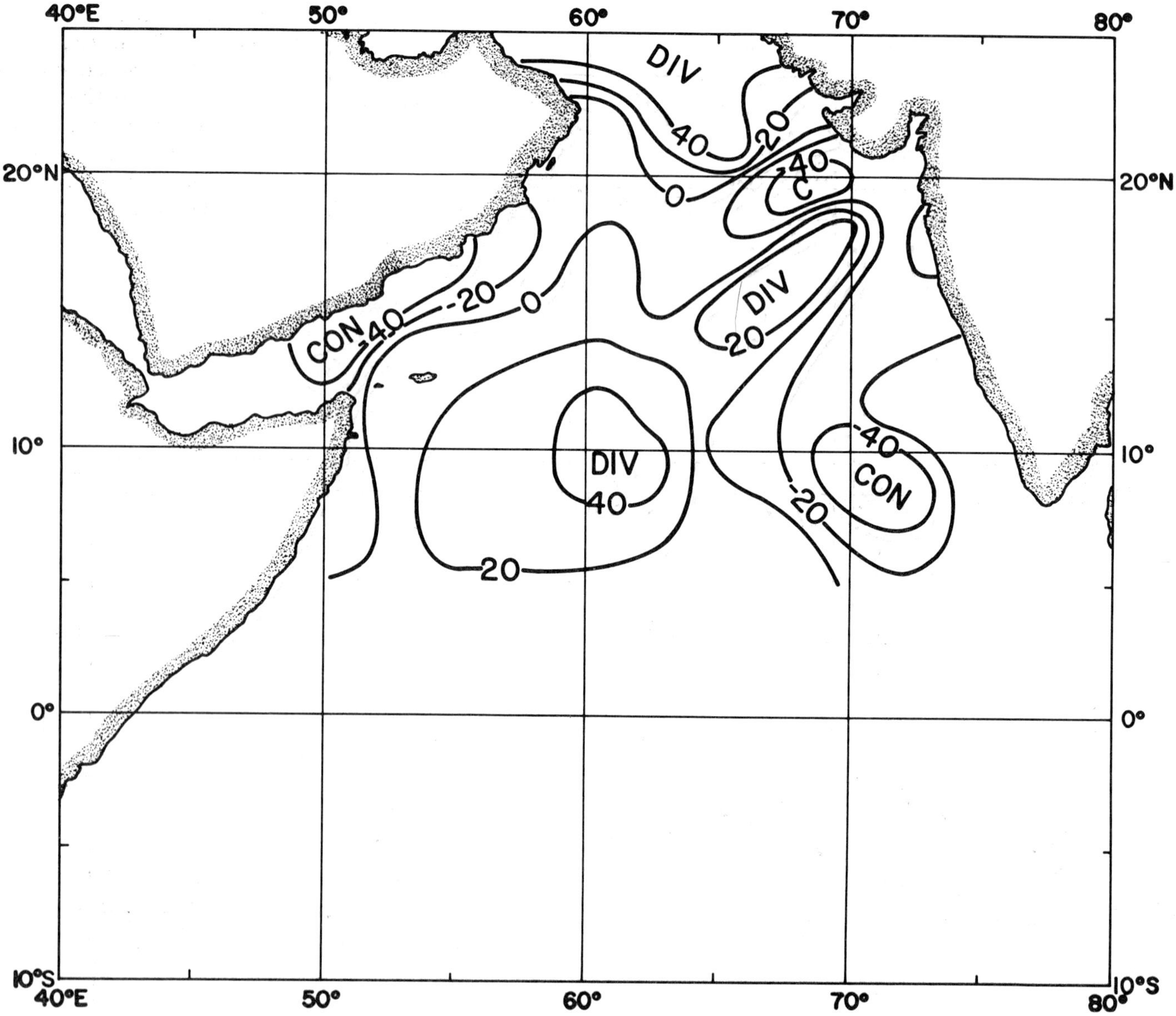

FIGURE 58. Computed divergence pattern for the synoptic case of 27 June 1963 over the Arabian Sea. Units: 10^{-6} sec^{-1}.

Both experimental trials of the numerical model for the frictional layer give promising results. Convergent regions are clearly to be identified with precipitation and divergent areas with a lack of precipitation. The locations may not be exact with the constant value of k selected, but the association between computed convergence and observed precipitation is most encouraging.

V. *Interpretation: The Divergence Equation*

After computing the horizontal divergence along selected trajectories, we examined the consistency of the results and tried to interpret the factors producing the results. We therefore examined the substantial derivative of the horizontal divergence. This naturally suggested an examination of the divergence equation obtained by taking the horizontal divergence of the vector momentum equation; the direct, but lengthy result is

$$\frac{dD}{dt} = \tfrac{1}{2}(\xi^2 - D^2 - \mathrm{Def}^2) - w\frac{\partial D}{\partial z} - \left(\frac{\partial w}{\partial x}\frac{\partial u}{\partial z} + \frac{\partial w}{\partial y}\frac{\partial v}{\partial z}\right) + \xi f - \beta u - \alpha \nabla^2 P - kD. \tag{99}$$

Further, it is easily verified that

$$\nabla_h \cdot \dot{\mathbf{V}}_h = \frac{dD}{dt} + \tfrac{1}{2}(D^2 - \xi^2 + \mathrm{Def}^2) + w\frac{\partial D}{\partial z} + \left(\frac{\partial w}{\partial x}\frac{\partial u}{\partial z} + \frac{\partial w}{\partial y}\frac{\partial v}{\partial z}\right), \tag{100}$$

that

$$\nabla_h \cdot (f\mathbf{V}_h \times \mathbf{K}) = \xi f - \beta u, \tag{101}$$

that

$$\nabla_h \cdot (\alpha \nabla_h P) = \alpha \nabla_h^2 P, \tag{102}$$

and that

$$\nabla_h \cdot (k\mathbf{V}_h) = kD. \tag{103}$$

In other words, in Equation (99) the first term on the left together with the first six terms on the right arise from taking the divergence of the horizontal acceleration; the seventh and eighth terms $(\xi f - \beta u)$, from taking the divergence of the horizontal Coriolis acceleration; the ninth and tenth terms $(-\alpha\nabla_h^2 P - kD)$ from taking the divergence of the pressure and frictional forces, respectively, or, briefly, that

$$\nabla_h \cdot \dot{\mathbf{V}}_h + \nabla_h \cdot (f\mathbf{V}_h \times \mathbf{K}) + \nabla_h \cdot (k\mathbf{V}_h) + \nabla_h \cdot (\alpha\nabla_h P) = 0. \tag{104}$$

We may refer to these four major contributions as acceleration, Coriolis, friction, and pressure-force terms respectively (*89*).

For horizontal motion, Equation (99) reduces to

$$\frac{dD}{dt} = \tfrac{1}{2}(\xi^2 - D^2 - \mathrm{Def}^2) + \xi f - \beta u - \alpha \nabla_h^2 P - kD. \tag{105}$$

Later, we consider the likely effect of ignoring the vertical motion terms, but first consider the physical content of Equation (105).

Considering the last two terms on the right-hand side of Equation (105), the frictional and pressure-force terms, it seems physically evident that a divergent force field is, in principle, capable of producing divergence in the fluid motion, that is, of contributing to the left-hand side of Equation (105). Moreover, we note that our formulation of friction leads to a term proportional to the divergence itself (kD) in (105). Thus, this term produces changes in the substantial derivative of the horizontal divergence (dD/dt) that tend to counteract the sign of the divergence itself—that is, if D is positive, dD/dt is negative, and conversely. This particular feature is a useful one in the present computations, since it acts effectively to restrain the vertical speed and thus does not aggravate our initial assumption of horizontal motion. Further, it is also useful to consider a more general combined result of divergent material and inertial accelerations and divergent force fields. In what follows, we examine our results in the light of both Equations (104) and (105) providing, on the one hand, a broad view of the four major terms and, on the other, a closer view at the composition of acceleration and Coriolis terms. Specifically, we tried to evaluate all terms in (105) save the deformation, which was obtained from elementary bookkeeping.

One serious consequence of ignoring the vertical-motion terms is to render the deformation negative, that is, imaginary; the discrepancies, generally of order 1×10^{-11} sec^{-2}, can usually be accounted for by including these terms, at least in an approximate way. This was the case for the Indian Ocean in July.

Remembering that our parcel is the locus of two or more sample points, we wished to evaluate the terms in (105) consistent with this description. For example, the relative vorticity was determined by evaluating twice the angular speed at the midpoint of a line connecting two sample points defining our parcel. The angular speed was computed by evaluating the change in the angle determined by the intersection of two straight lines connecting two sample parcels on adjacent trajectories at two different times; twice the angular speed is just the relative vorticity. We adopted the familiar sign convection for the relative vorticity and assigned local clockwise rotation a negative sign. Again, the Coriolis parameter (f) appearing in the vorticity term (ξf) was also evaluated at the same midpoint, as well as the remaining part (βu) of the Coriolis term.

On the other hand, the evaluation of the pressure-force term was not quite so obvious. In the first place, this term ($\alpha \nabla_h^2 P$) is basically a Eulerian property of the pressure field, and would normally be evaluated by, say, finite difference techniques applied directly to a pressure map. Though the question of a suitable horizontal distance might be bothersome, this too could be decided in various plausible ways. The real question, however, was whether we wished to evaluate some terms of Equation (105) "following the motion" and to use a field description in the case of the pressure-force term. Greater consistency is served by adopting the former Lagrangian view and appealing to the following intuitive argument: since we knew the pressure distribution in space, and the distance between sample points (s) defining our fluid parcel, we could compute the Laplacian using the familiar finite difference expression

$$\nabla_h^2 P = \frac{P_1 + P_2 - 2P_0}{(\delta s)^2}, \tag{106}$$

knowing the value of the pressure (P_1, P_2) at the extremities of the fluid parcel, and at the midpoint (P_0). This procedure assured us that the pressure-force term was evaluated in a manner consistent with that adopted for the other terms, and avoided the nagging doubts entailed in attempting to use a single estimate of the distance, s, throughout the entire trajectory.

As a sample test, we evaluated the Laplacian using both methods for the mean pressure distribution prevailing over the Indian Ocean during July. In the direct field evaluation, we took s to be 2.5 degrees of latitude everywhere, while in the Lagrangian evaluation, s varied between 2 and 4 degrees of latitude along one sample trajectory. The resulting patterns were quite similar, though because s necessarily varied along the trajectory in the Lagrangian case, actual values differed by as much as a factor of two in some cases.

Some appeal could be made to Gerber's (*90*) development of the Laplacian, though the advantages to the present investigation are not obvious.

A. CASE STUDIES

1. Indian Ocean—January

What are the factors responsible for producing the computed divergence along the trajectories of Figure 28? Since the distribution of divergence was basically zonal, it is sufficient to exhibit the results for just two sample parcels; we chose the two parcels initially located at 10° and 15°N (indicated by asterisks in Figure 29) and examined the results in the light of Equations (104) and (105), that is, how do these results satisfy the divergence equation?

First, we considered the results in the context of Equation (104), displaying the four major contributions: the divergence of acceleration, pressure force, friction, and Coriolis terms (Figure 59)—then in terms of (105) wherein the acceleration and Coriolis terms are displayed explicitly (Figure 60). We wanted to understand what factors determine curve D (Figure 59) and the left-hand side of (105), plotted as squares in Figure 62.

Between about 9.0°N and 0.3°S (Figure 59) the divergence of the acceleration was nearly equal but opposite to the divergence of the Coriolis terms, with the divergence of the frictional term—here proportional to the divergence itself, and briefly referred to as the divergence—being determined by the divergence of the pressure-force term; since the latter was determined from the pressure field the divergence itself can be inferred and, by Equation (105), must be negative in the present case. Further, there was a lag between the divergence of the pressure-force term and the divergence itself; this is understandable if we recognize that some finite time (about 10 hours in the present case) is required for an air parcel to adjust its motion to a spatial change in the pressure gradient. As a result, such changes are displaced downstream relative to changes in the pressure gradient. (Changes in the latter quantity near 5°N amounted to 0.4 mb in 5.1° latitude.) Further, the small area of positive divergence near 8°N was essentially the consequence of the divergence of the Coriolis terms, since this quantity remained negative even after the acceleration and pressure-force terms were subtracted (the divergence must remain positive).

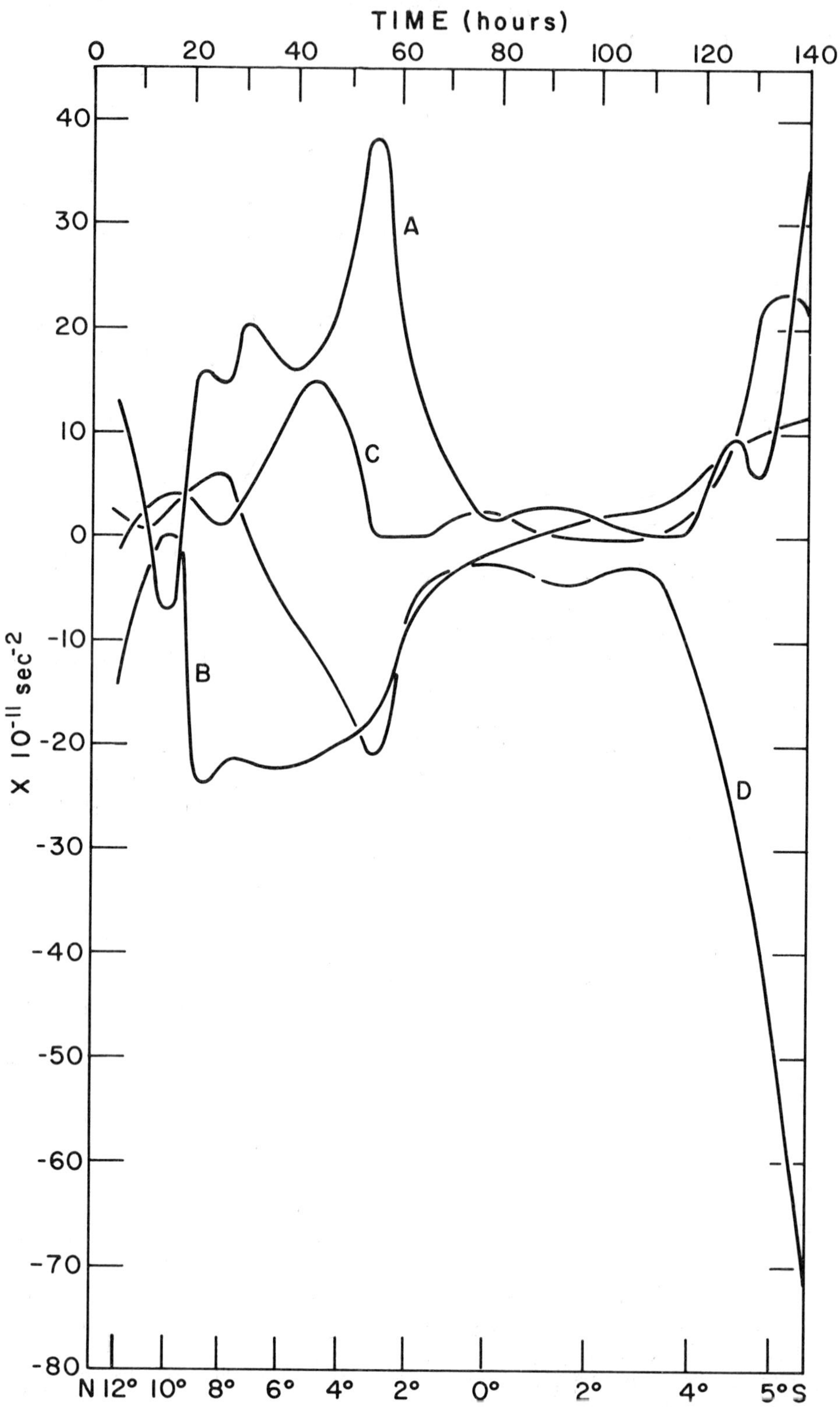

FIGURE 59. The divergence of the acceleration (A), Coriolis forces (B), pressure-gradient force (C), and frictional force (D), for the two sample parcels indicated by the asterisks in Figure 29. The values are taken from Table 4; the plotted curves satisfy Equation (104).

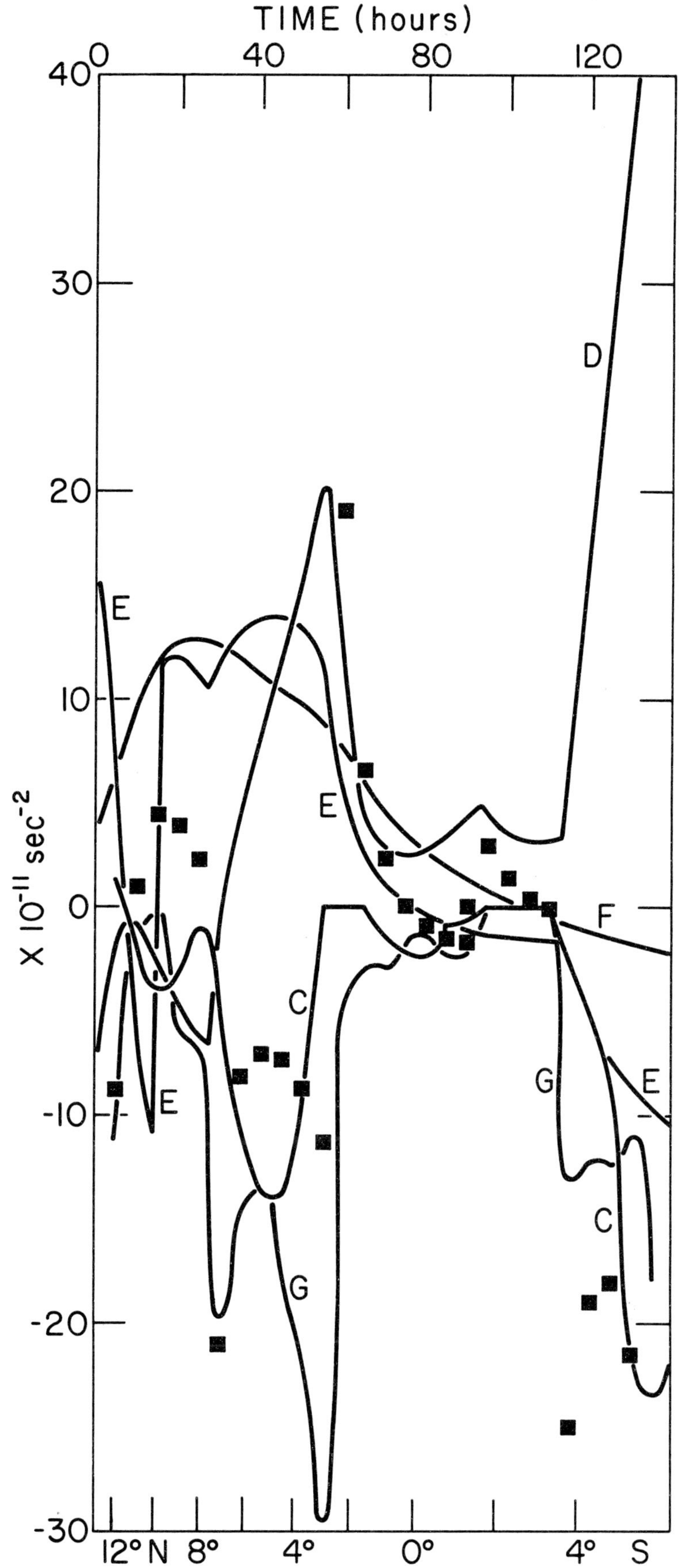

FIGURE 60. An expanded form of Figure 59 showing the explicit contributions to the divergence of the Coriolis terms: βu, Curve F, and ξf, Curve E; two of the four terms comprising the divergence of the acceleration: dD/dt, solid squares, ½Def², Curve G. The remaining curves, C and D, are the same ones appearing in Figure 61. Units are × 10^{-11} sec^{-2}; the curves have been plotted with due regard for the algebraic signs in Equation (104).

Southward from 3°N the divergence decreased rather rapidly—largely the result of the rapid decrease in the divergence of the pressure-force term. Between the equator and 3°S the divergence itself was negative but small (since the other terms had small positive values and all ordinates must add to zero). Again, south of 3°S, all terms, save the divergence, increased, thus demanding a decrease in the latter to a value of about -40×10^{-6} sec^{-1} near 5.5°S.

In contrast to (104), we exhibit the same result in terms of (105) (Figure 60). Again, all terms of (105) were evaluated save the deformation term, which was deduced from elementary bookkeeping. The results for the same two parcels are displayed in Figure 60. Here the two curves D and C are the same that appeared in Figure 59 with the addition of two explicit curves (E and F) for the Coriolis terms and the time derivative of the horizontal divergence plotted as discrete squares rather than a continuous curve, to avoid clutter on the diagram. All values, it should be noted, were computed in hourly time-steps with Table 4 containing the numerical values at 5-hour intervals.

Recalling that Figure 60 was plotted with due regard for the algebraic signs appearing in (105), we see that the two terms comprising the divergence of the Coriolis forces were negatively correlated north and south of the equator, and positively correlated within the narrow region 0.2°S – 1.3°S. Broadly speaking, both terms were about the same order of magnitude though βu was slightly larger within a degree or so of the equator.

The divergence ($-D^2/2$), deformation ($-\text{Def}^2/2$), and friction ($-kD$) terms decrease the time derivative of the horizontal divergence; on the other hand, the vorticity term ($\xi^2/2$) increases it, while the Coriolis ($\xi f - \beta u$) and pressure-force ($-\alpha \nabla_h^2 P$) terms may act in either sense. Generally speaking, both the divergence ($-D^2/2$) and vorticity ($\xi^2/2$) terms are usually smaller than the other terms mentioned above.

Again, near 10°N the positive contribution to the time derivative of the horizontal divergence, and the resulting increase in the divergence itself, was largely the effect of the Coriolis terms; specifically the βu term up to about 10°N when this was reinforced by the ξf term. The large oscillations in the latter term between 11° and 12°N are not entirely clear; perhaps this represents the initial adjustment to the coarse initial state and, later, the adjustment to the rapid change in the divergence of the pressure-force term. In any event, both Coriolis terms contribute positively to a local divergence extremum (positive) near 7.5°N; by contrast, the pressure-force term acted in the opposite direction. Near 7.5°N the derivative of the divergence changed sign and remained negative, producing a minimum in the divergence itself near 3°N; during this time, the divergence of the pressure-force term increased to a maximum (of 14.0×10^{-11} sec^{-2}), while the deformation was nearly, but not quite, balanced by the Coriolis terms. Thus, near 4.5°N, where the pressure-force term was a maximum, it made the largest single contribution to decreasing the parcel divergence along the motion; the deformation made the second. As the divergence of the pressure-force term decreased to zero near 2.5°N, the deformation became the largest single contributor to decreasing the divergence itself, though, because we neglected the vertical-motion terms in (105), the deformation is likely to be an overestimate. By the time the parcel reached 1.5°N, the divergence of the pressure-force had vanished and the major positive contributions to the time derivative of the divergence arose largely from the vorticity ($\frac{1}{2}\xi^2$)

and friction ($-kD$) terms, since again the Coriolis and deformation terms nearly cancelled. Near 1.5°S the divergence attained a small relative minimum, again largely the result of the pressure-force term and, as before, displaced downstream by about 2 degrees of latitude (equivalent to 10 or 12 hours). Finally, near 3.5°S all terms, save the frictional term ($-kD$), produced negative contributions to the time derivative of the divergence leading to a rapid decrease in the divergence itself to about -40×10^{-6} sec^{-1} near 5.5°S.

The strong convergence near 5°S also agrees with the seasonal character of the rainfall at the Seychelles (5°S, 55°E), namely, a maximum. However, we must acknowledge that a similar result is likely with a weak trough between 5–10°S, rather than zero gradient as we have used; unfortunately, the pressure data of the Dutch atlas lacks sufficient detail to depict such a feature if it exists.

To summarize both descriptions embodied in Figures 59 and 60 in

TABLE 4. Computed values ($\times 10^{-11}$ sec^{-2}) of terms in Equation (105) for the 2 parcels indicated by asterisks in Figure 29, plus the divergences of the acceleration (A), Coriolis force (B), pressure-gradient force (C), and friction (D); the former are plotted in Figure 60, the latter in Figure 59. Indian Ocean, January.

Time	Latitude	$\frac{dD}{dt}$	$-\frac{1}{2}D^2$	$\frac{1}{2}\xi^2$	$-\frac{1}{2}Def^2$	$\nabla \cdot \dot{V}_H$ (A)	$f\xi$	$-\beta u$	$\nabla \cdot (fV_H)$ (B)	$-kD$ (D)	$-\alpha\nabla^2 P$ (C)	$\frac{\partial w}{\partial y}\frac{\partial v}{\partial z}$
5	11.9°N	−8.8	0	0.3	−21.9	12.8	6.8	7.0	13.8	−2.0	1.0	−
10	11.0	1.0	0	0.1	−1.8	2.7	−4.5	9.0	5.5	−0.8	−2.0	+
15	10.0	4.4	0	0.9	−0.3	3.8	−10.7	10.4	−0.3	−2.6	−4.1	+
20	9.1	3.9	−0.1	1.4	−12.4	15.0	12.1	10.9	23.0	−4.7	−3.3	+
25	8.1	2.3	−0.2	1.5	−13.7	14.7	11.1	10.9	22.0	−6.2	−1.1	−
30	7.0	−21.1	0	2.0	−39.5	20.4	11.3	10.4	21.7	−0.9	−4.4	−
35	6.2	−8.2	−0.1	3.5	−29.2	17.6	12.9	9.3	22.2	4.3	−8.9	−
40	5.3	−7.1	−0.4	4.8	−27.5	16.0	13.7	8.1	21.8	8.0	−13.8	−
45	4.6	−7.4	−0.8	9.1	−34.4	18.7	13.9	7.2	21.1	11.6	−14.0	−
50	3.8	−8.7	−1.5	8.8	−42.1	26.1	13.3	6.3	19.6	15.7	−9.2	−
55	2.9	−11.4	−2.8	12.0	−58.8	38.2	11.8	5.6	17.4	20.8	0	+
60	2.2	19.2	−0.8	7.2	−9.1	21.9	6.1	4.6	10.7	11.2	0	+
65	1.5	6.5	−0.2	2.4	−6.6	10.9	2.6	3.5	6.1	4.8	0	+
70	0.9	2.3	0	1.0	−6.5	7.8	1.1	2.5	3.6	3.0	−1.4	+
75	0.3	0.1	0	0.6	−2.9	2.4	0.3	1.8	2.1	2.5	−2.2	−
80	0.2°S	−1.0	−0.1	0.5	−3.2	1.8	−0.2	1.5	1.3	2.8	−2.3	−
85	0.8	−1.5	−0.1	0.7	−4.8	2.7	−0.6	0.6	0	3.5	−0.8	−
90	1.3	−1.7	−0.1	0.3	−4.6	2.7	−1.0	0.2	−0.8	4.3	−0.8	−
95	1.8	2.9	−0.1	0.3	−0.2	2.9	−1.3	−0.3	−1.6	4.5	0	+
100	2.4	1.3	−0.1	0.3	−0.4	1.5	−1.4	−0.7	−2.1	3.6	0	+
105	2.9	0.3	−0.1	0.2	−0.5	0.7	−1.5	−1.0	−2.5	3.2	0	0
110	3.3	−0.2	−0.1	0.2	−0.6	0.3	−1.6	−1.3	−2.9	3.2	0	−
115	3.8	−25.1	−0.3	0.4	−25.5	0.3	−2.8	−1.6	−4.4	7.2	−2.2	−
120	4.2	−19.2	−2.0	1.4	−24.2	5.6	−5.5	−1.6	−7.1	17.8	−5.1	−
125	4.5	−18.2	−4.5	2.0	−25.2	9.5	−7.3	−1.5	−8.8	27.0	−8.7	−
130	4.8	−21.5	−8.8	2.6	−21.2	5.9	−8.6	−1.4	−10.0	37.0	−21.1	−
135	5.0	−32.4	−16.4	2.9	−35.1	16.2	−9.7	−1.3	−11.0	50.7	−23.5	−
140	5.2	−64.5	−36.1	3.4	−73.8	42.0	−10.6	−1.3	−11.9	75.3	−21.4	−
145	5.4	−34.4	−87.4	2.4	−55.6	106.2	−9.6	−1.2	−10.8	117.0	0	−

terms of the relative extrema in the horizontal divergence (Curve D) we recall that both extrema near 11° and 9.5°N were the result of the divergence of the Coriolis terms (βu in the first case, $f\xi$ and βu in the second); the maximum convergence near 2.5°N and the modest one near 1.5°S resulted from the divergence of the pressure-force term, while, lastly, the large convergence near 5°S resulted from three of the four terms in Equation (104) acting in the same sense. Or, more succinctly, the distribution of horizontal divergence within 5 degrees of the equator is largely determined by the divergence of the pressure-force term, while poleward of these latitudes the Coriolis terms are dominant.

2. Indian Ocean—July

Again, it will be sufficient to examine the results of two sample parcels (indicated by asterisks in Figure 33); in particular, those parcels initially located at 15°N and 20°N respectively (Table *5*).

TABLE 5. Computed values ($\times\ 10^{-11}\ \text{sec}^{-2}$) of terms in Equation (105) for the 2 parcels indicated by asterisks in Figure 33, plus the divergences of the acceleration (A), Coriolis force (B), pressure-gradient force (C), and friction (D); the former are plotted in Figure 62, the latter in Figure 61. Indian Ocean, July.

Time	Latitude	$\frac{dD}{dt}$	$-\frac{1}{2}D^2$	$\frac{1}{2}\xi^2$	$-\frac{1}{2}\text{Def}^2$	$\nabla\cdot\dot{V}_H$ (A)	$f\xi$	$-\beta u$	$\nabla\cdot(fV)$ (B)	$-kD$ (D)	$-\alpha\nabla^2P$ (C)	$\frac{\partial w}{\partial y}\frac{\partial v}{\partial z}$
5	15.3°S	7.5	−0.2	0.6	−31.4	−38.5	14.5	15.6	30.1	−4.7	13.1	+
10	15.0	8.3	−0.5	0.7	−29.8	−37.9	15.3	19.5	34.8	−8.9	12.0	+
15	13.6	5.1	−0.9	0.5	−0.6	−6.1	10.9	21.1	32.0	−12.2	−13.7	+
20	11.6	−31.1	−0.5	0.5	−47.3	−16.2	9.3	21.4	30.7	−8.7	−5.8	−
25	10.8	−10.4	0	1.8	−45.0	−32.8	16.9	21.0	37.9	−0.4	−4.7	−
30	9.6	−6.1	0	4.4	−43.1	−32.6	22.6	20.2	42.8	3.4	−13.6	−
35	9.2	−4.8	0	8.1	−43.5	−30.6	25.4	19.6	45.0	6.0	−20.4	−
40	6.9	−22.0	−1.4	13.3	−80.3	−46.4	29.0	19.1	48.1	14.7	−16.4	+
45	5.7	11.4	−0.5	10.1	−32.0	−33.8	21.3	17.0	38.3	8.9	−13.4	+
50	4.6	2.2	−0.2	7.0	−29.0	−24.5	13.4	13.9	27.3	6.1	−8.9	+
55	3.6	−3.4	−0.3	6.5	−33.2	−23.6	8.8	11.3	20.1	6.8	−3.3	−
60	2.5	−6.6	−0.6	4.2	−32.7	−22.5	5.8	9.2	15.0	9.6	−2.1	+
65	1.4	−3.2	−0.3	2.3	−22.1	−16.9	2.6	7.0	9.6	7.3	0	−
70	0.1	−5.8	−0.6	2.5	−23.2	−15.6	0.9	4.9	5.8	9.8	0	−
75	0.3°N	−7.5	−1.1	0.5	−22.8	−15.9	−0.4	3.1	2.7	13.2	0	−
80	1.1	−9.1	−2.0	2.5	−26.8	−17.2	−1.8	1.5	−0.3	17.5	0	−
85	1.8	0.4	−3.0	1.7	−17.6	−18.5	−3.0	0	−3.0	21.5	0	+
90	2.5	8.2	−1.7	1.0	−3.7	−12.6	−2.8	−1.2	−4.0	16.6	0	+
95	3.1	5.3	−1.1	0.5	−5.7	−11.6	−2.5	−2.2	−4.7	13.4	2.9	+
100	3.6	3.4	−0.8	0.4	−7.4	−11.2	−2.2	−3.0	−5.2	11.4	5.0	+
105	4.1	2.2	−0.6	0.2	−11.5	−14.1	−2.0	−3.5	−5.5	10.0	9.6	+
110	4.6	1.4	−0.5	0.1	−8.0	−9.8	−1.9	−4.0	−5.9	9.2	6.5	+
115	5.0	1.0	−0.5	0.1	−2.3	−3.7	−1.8	−4.3	−6.1	8.6	1.2	+
120	5.4	0.7	−0.4	0.1	−6.4	−7.4	−1.7	−4.5	−6.2	8.2	5.4	+
130	6.1	0.5	−0.4	0	−11.6	−12.5	−1.6	−4.9	−6.5	7.6	5.9	+
140	6.8	0.5	−0.3	0	0	−0.7	−1.4	−5.0	−6.4	7.1	11.4	+
150	7.4	0.5	−0.3	0	0	−0.1	−1.3	−5.2	−6.5	6.6	1.8	+
160	8.0	0.4	−0.3	0	0	0.3	−1.2	−5.2	−6.4	6.1	0	+
170	8.6	0.4	−0.2	0	0	0.6	−1.1	−5.2	−6.3	5.7	0	+
180	9.1	0.4	−0.2	0	0	0.9	−1.0	−5.2	−6.2	5.3	0	+

Again, we present our analysis in the context of Equation (104) in Figure 61, and in the expanded form of Equation (105) in Figure 62. We recall that curves B, C, and D (Figure 61) were determined independently, while A is the algebraic sum of B, C, and D subject to the additional control that three of the four terms (dD/dt, $D^2/2$, $\xi^2/2$) comprising A, the divergence of the acceleration, could also be computed. The fourth, the deformation ($\text{Def}^2/2$), could be computed by elementary bookkeeping. Ultimately, curve A is valid to the extent that the vertical motion terms in (105) could be neglected.

Referring to Figure 62, we see that near 7°S the Coriolis and acceleration terms nearly balanced each other with the resulting divergence being determined by the pressure-force term; this led to a relative maximum of the divergence near 14°S, lagging the minimum in the pressure-force term by about 1.5 degrees of latitude. At this point the acceleration term (Curve A) decreased rapidly, the pressure-force term (Curve C) became positive, and since the Coriolis term (Curve B) had changed little, the divergence itself (Curve D), by (104), decreased. This decrease (Curve D) was sustained until the divergence became negative, reaching a relative minimum near 7°S. The sustained decrease was largely the result of the pressure-force term, since both acceleration and Coriolis terms nearly cancelled between 5 and 10°S.

Again, we see the familiar lag in the divergence extrema between the pressure-force term and the divergence (Curves C and D), the latter appearing some 1.5 degrees downstream, slightly less than observed in the January case since the July speeds were slightly higher.

Subsequently, the divergence (Curve D) increased to a relative maximum (but still remained negative) near 5°S, since here, though Coriolis and acceleration terms were roughly balanced (Curves A and B), the pressure-force term (Curve C) was decreasing, and again, by (104), the divergence increased. Further along, the divergence decreased again reaching a minimum value near 2°N; now, however, the pressure-force term was zero and the Coriolis term was increasing. Thus for the first time the acceleration term determined the resulting divergence. (It will be noted below that the largest single contribution to the acceleration term comes from the deformation.) Northward of 2°N the pressure-force term (Curve C) and the acceleration term (Curve A) nearly balanced, and with the Coriolis term being positive and nearly constant, the divergence, by (104), was negative and slowly increasing. This feature, the near balance between acceleration and pressure-force terms, was also observed near 16°S, and would be expected near 4°N in the light of (104) and the near balance northward of 4°N between Coriolis and frictional terms. The July case, unlike January, contained some negative values of the pressure-force terms.

An equivalent analysis in the context of (105) serves to indicate the specific composition of the acceleration and Coriolis terms, and how they contributed to the temporal change of the horizontal divergence, as well as the pressure-force and frictional terms discussed above.

Table 5 and Figure 62 show that the Coriolis terms (ξf, βu), with the exception of a small region just north of the equator (.2–1.8°N), were oppositely correlated throughout; south of the equator both terms were of about the same order of magnitude—with some obvious exceptions—while north of the equator βu was slightly larger than ξf.

Between 15° and 4.6°S the largest contribution to the divergence of the acceleration came from the deformation term, and the second largest

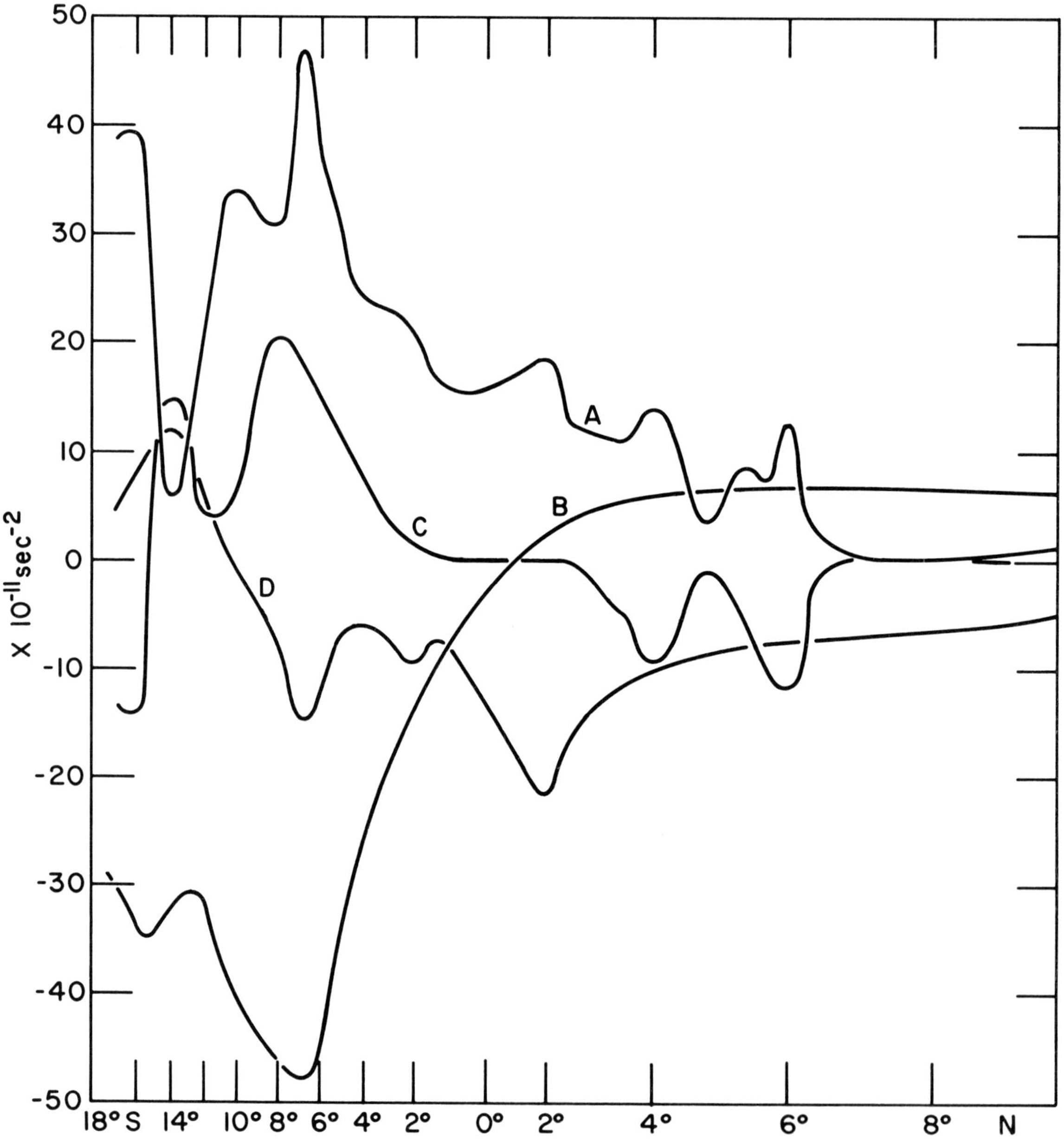

FIGURE 61. The divergence of the acceleration (A), Coriolis forces (B), pressure-gradient force (C), and frictional force (D) for the two sample parcels indicated by the asterisks in Figure 33. The numerical values ($\times 10^{-11}$ sec^{-2}) are taken from Table 5; the plotted curves satisfy Equation (104).

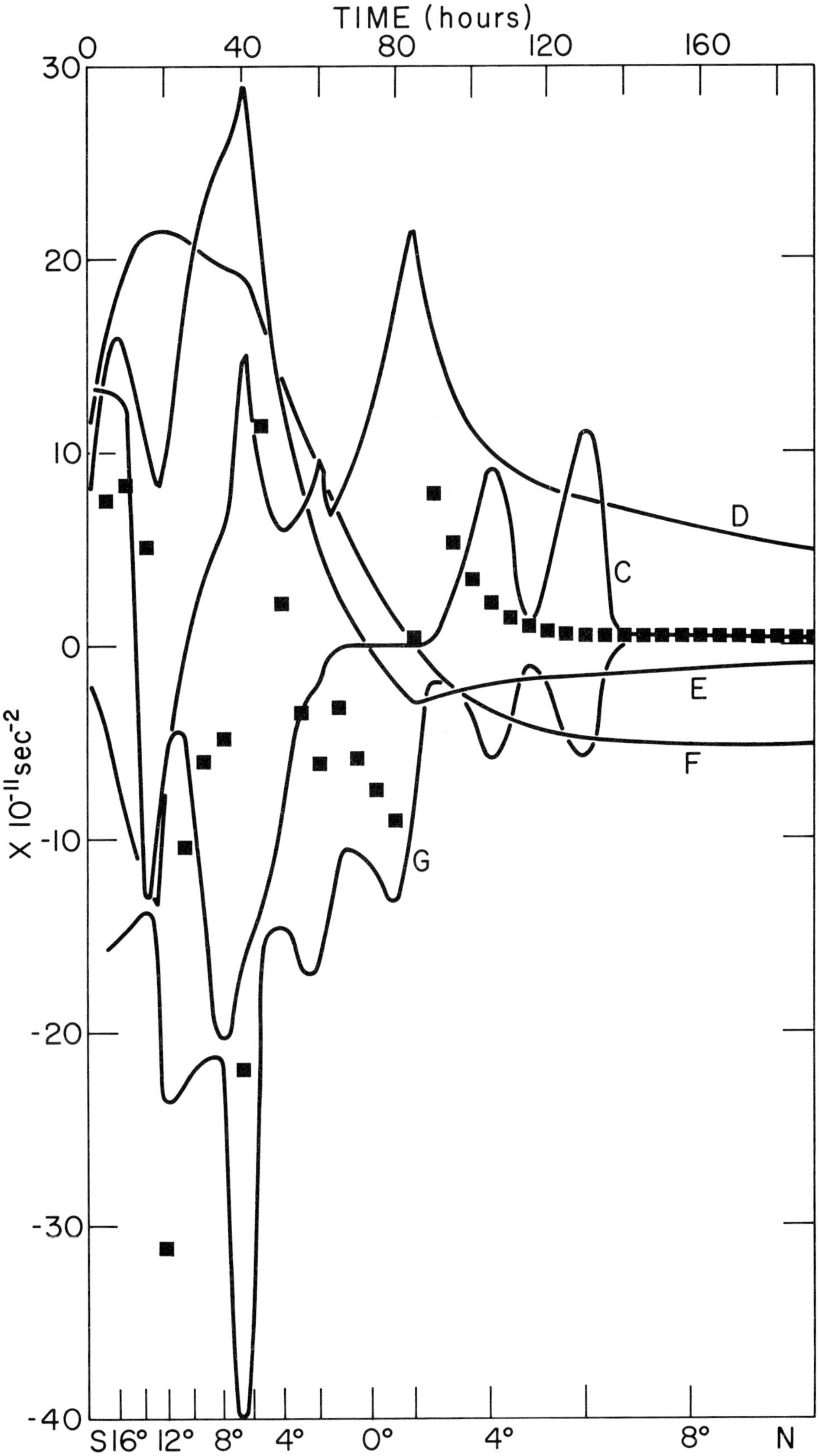

FIGURE 62. An expanded form of Figure 61 showing the explicit contributions to the divergence of the Coriolis terms: βu, Curve F, and ξf, Curve E; two of the four terms comprising the divergence of the accelerations: dD/dt, solid squares, $\frac{1}{2}Def^2$, Curve G. The remaining curves, C and D, are the same ones appearing in Figure 61. Units are $\times 10^{-11}$ sec^{-2}; the curves have been plotted with due regard for the algebraic signs in Equation (104).

from the time derivative of the horizontal divergence. This term was obtained from our computations of horizontal divergence, and then the deformation was derived from elementary bookkeeping through a neglect of the vertical motion term in (105). The neglect of the latter term can express itself practically by leading to small negative quantities—that is, imaginary—for the deformation; this is, in fact, the case for all computations of the deformation between 6.8° and 10°N. However, by computing the vertical motion, through a 1 km layer, from the available computations of divergence, and neglecting zonal variations, in the vertical motion, we found that for a vertical wind shear of 2 m sec^{-1} km^{-1} the vertical motion term is about $+0.5 \times 10^{-11}$ sec^{-1} near 8°N. This gave zero deformation at 7.1°N; accordingly, the deformation northward of 6.8°N was assumed to vanish.

Though no imaginary deformations were computed between 15°S and 6.5°N, it is useful to see what general effect the neglect of the vertical motion is likely to have. By assuming the east-west variations in the vertical motion are smaller than the north-south ones, and that the vertical wind shear is about 1 or 2×10^{-3} sec^{-1}, then the sixth term of Equation (100) varies between 1 and 4 in the region 15°S to 6.5°N. Specifically, at 15°S it was about 2.4×10^{-11} sec^{-2}, and thus the deformation was overestimated by about 8%; near 7.3°N where the vertical-motion term was about 1.3×10^{-11} sec^{-2}, the deformation was underestimated by about 4%. Since the estimated error never exceeded 10% no corrections were applied to the derived deformations between 15°S and 6.5°N.

The increase in horizontal divergence along the trajectory near 5°S (Figure 33) is not without some climatological interest occurring, as it does, in the latitudes of the Seychelles; here, surprisingly, the mean July rainfall is a relative minimum (*91*). Attempts to relate this feature to mean trough positions have not been successful (*80*). In the context of Equation (105), our computations show the increase in horizontal divergence near 5°S to be the result of a decreasing divergent pressure-force term (Figure 61), since the divergence of the Coriolis and acceleration terms nearly balance in this region: thus the kD term must increase.

To summarize then, the distribution of horizontal divergence along our sample trajectory between 20°S and 10°N is a consequence of: (*a*) a divergent pressure force between 20°S and the equator, and (*b*) divergent acceleration (0–4°N) and Coriolis terms (north of 4°N).

3. Gilbert Islands—July 1964

Returning to Figure 39, how can we account for the rapid change in horizontal divergence between, say, the equator and 5°N? Specifically, what factors are responsible for altering the divergence along the trajectories formed by parcels released at the equator, 2°N, and 4°N?

Again, we based our analysis upon Equation (104) and present separate results for parcels released at the equator and 2°N (Figure 63) on the one hand and parcels released at 2°N and 4°N (Figure 64) on the other. The three parcels are identified by asterisks in Figure 38, while Tables 6 and 7 contain the relevant computations. How then are we to understand the relative shapes of the D curves in Figure 63 and 64?

We recall that the divergence of the acceleration (92) consists of four separate terms when the vertical-motion terms are ignored. According to Table 6, the divergence squared and the vorticity squared nearly cancelled each other, while the deformation squared was nearly an order of magnitude smaller than the remaining terms.

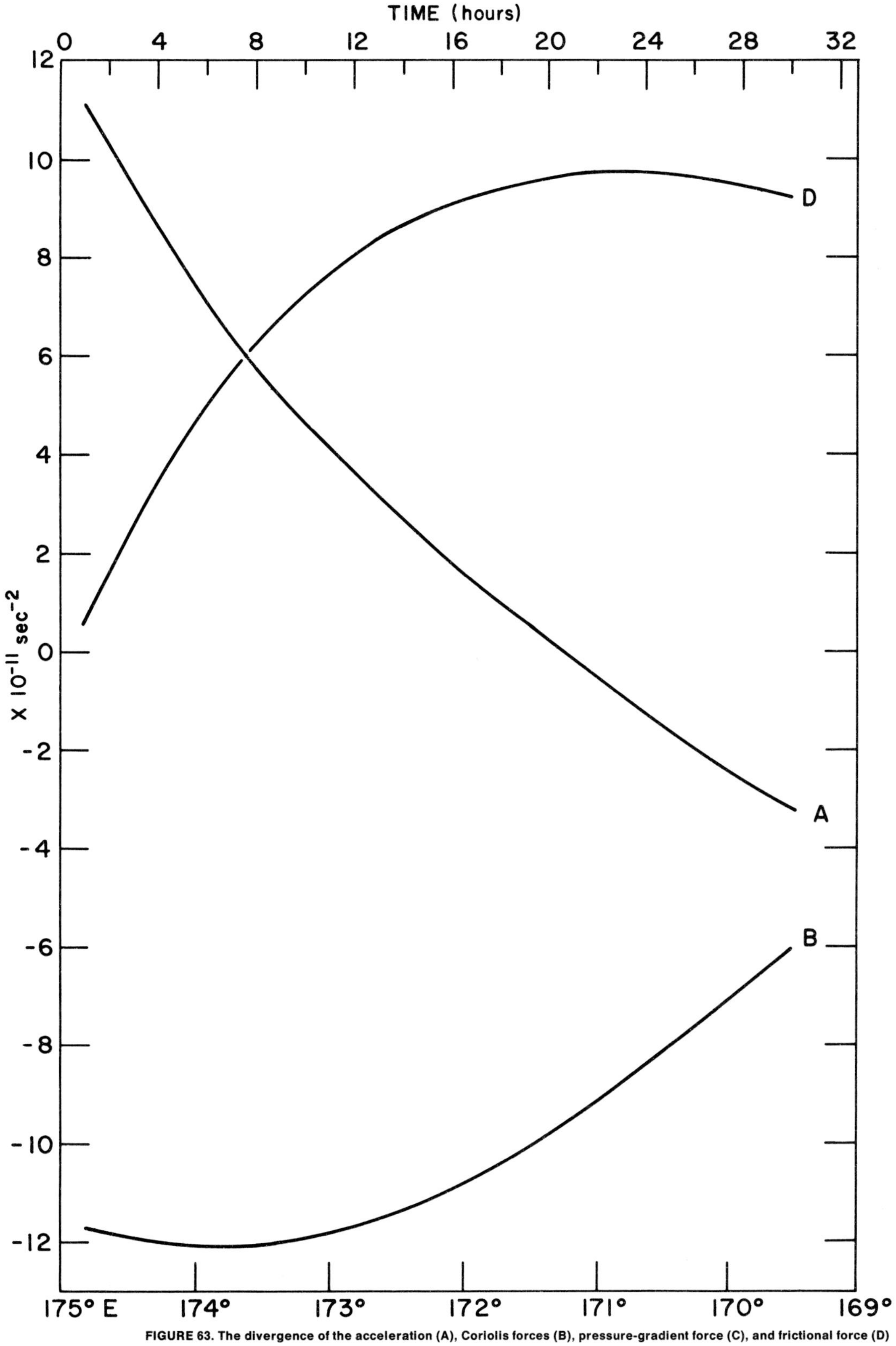

FIGURE 63. The divergence of the acceleration (A), Coriolis forces (B), pressure-gradient force (C), and frictional force (D) for the two parcels released at 0° and 2°N and indicated by asterisks on the right-hand margin of Figure 38. The numerical values are taken from Table 6 and satisfy Equation (104).

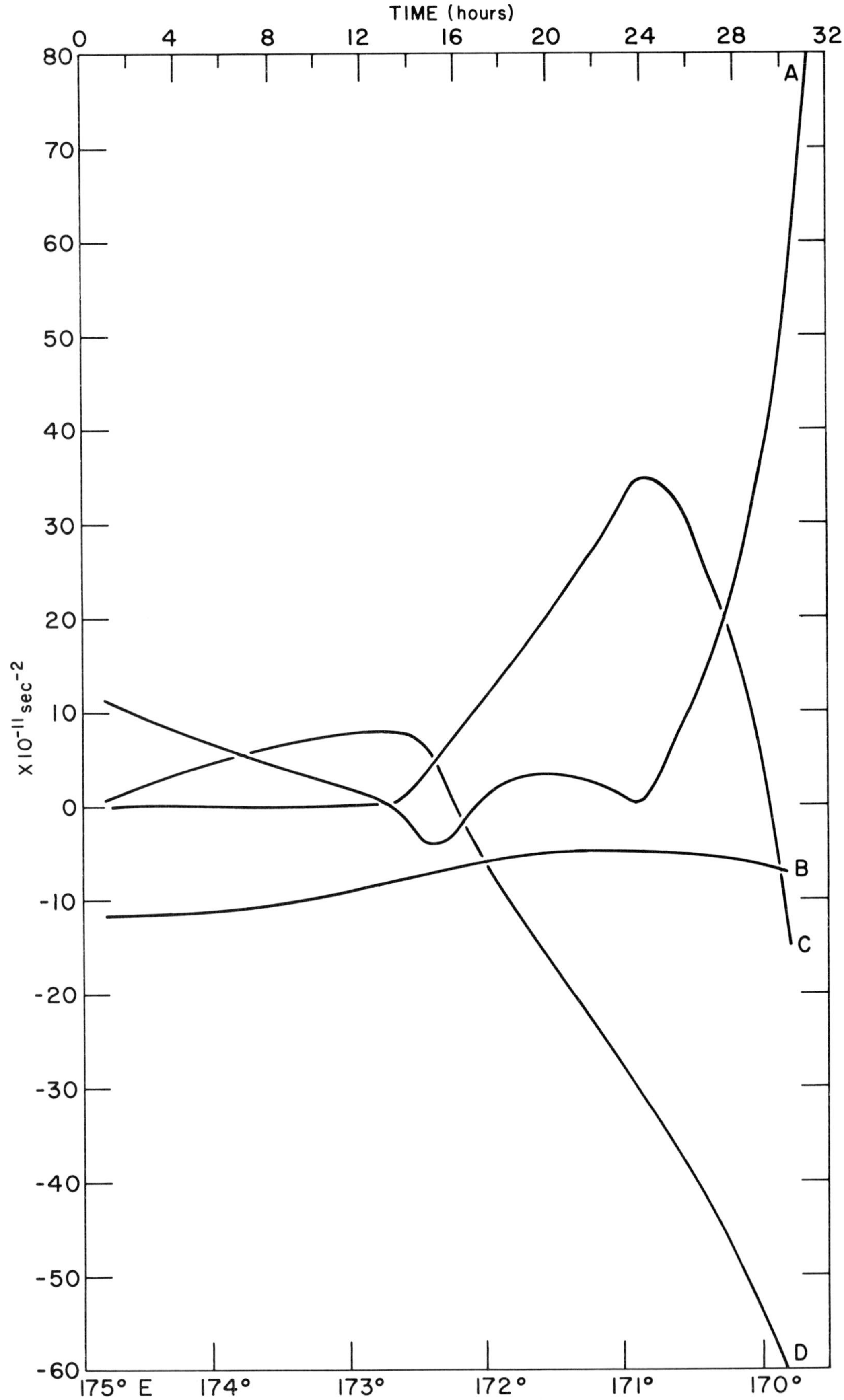

FIGURE 64. Same analysis as that presented in Figure 63, but for the two parcels released at 2° and 4°N. The numerical values are taken from Table 7 and satisfy Equation (104).

TABLE 6. Computed values ($\times 10^{-11}$ sec^{-2}) of terms in Equation (105) for the 2 parcels leaving 0° and 2°N in Figure 38, plus the divergences of the acceleration (A), Coriolis force (B), pressure-gradient force (C), and friction (D); the latter are plotted in Figure 63. Gilbert-Ellice Islands, July 1964.

Time	Long. E	$\frac{dD}{dt}$	$-\frac{1}{2}D^2$	$-\frac{1}{2}\xi^2$	$-\frac{1}{2}Def^2$	$\nabla \cdot \dot{V}_H$ (A)	$f\xi$	$-\beta u$	$\nabla \cdot (fV)$ (B)	$\alpha \nabla^2 P$ (C)	$-kD$ (D)	$\frac{\partial w}{\partial x}\frac{\partial u}{\partial z} + \frac{\partial w}{\partial y}\frac{\partial v}{\partial z}$
2	174.6	10.5	0	0	−0.3	10.2	0	11.8	11.8	0	−1.6	
4	174.4	8.9	−0.1	0	−0.5	8.5	−0.1	12.1	12.0	0	−3.5	+
6	174.0	7.3	−0.2	0	−0.4	7.1	−0.2	12.3	12.1	0	−5.0	
8	173.6	5.9	−0.2	0.1	−0.3	5.7	−0.4	12.4	12.0	0	−6.3	+
10	173.2	4.7	−0.4	0.2	−0.2	4.6	−0.6	12.5	11.9	0	−7.3	
12	172.8	3.6	−0.4	0.3	−0.1	3.6	−0.8	12.5	11.7	0	−8.1	+
14	172.5	2.8	−0.5	0.3	−0.3	2.6	−1.1	12.4	11.3	0	−8.7	
16	172.1	2.0	−0.5	0.4	−0.4	1.7	−1.5	12.3	10.8	0	−9.1	+
18	171.7	1.3	−0.5	0.3	−0.2	1.0	−1.8	12.2	10.4	0	−9.4	
20	171.4	0.8	−0.6	0.7	−0.2	0.2	−2.2	12.0	9.8	0	−9.6	0
22	171.1	0.3	−0.6	0.9	−0.3	−0.2	−2.7	11.8	9.1	0	−9.7	
24	170.8	−0.1	−0.6	1.0	−0.5	−0.6	−3.1	11.6	8.5	0	−9.7	+
26	170.4	−0.5	−0.6	1.2	−0.8	−1.2	−3.6	11.3	7.6	0	−9.6	
28	170.1	−0.8	−0.5	1.3	−1.0	−1.7	−4.2	11.0	6.8	0	−9.4	+
30	169.7	−1.1	−0.5	1.4	−1.2	−2.0	−4.7	10.7	6.0	0	−9.2	

TABLE 7. Computed values ($\times 10^{-11}$ sec^{-2}) of terms in Equation (105) for the 2 parcels leaving 2° and 4°N in Figure 38, plus the divergences of the acceleration (A), Coriolis force (B), pressure-gradient force (C), and friction (D); the latter are plotted in Figure 64. Gilbert-Ellice Islands, July 1964.

Time	Long. E $\frac{dD}{dt}$	$-\frac{1}{2}D^2$	$\frac{1}{2}\xi^2$	$-\frac{1}{2}Def^2$	$\nabla \cdot \dot{V}_H$ (A)	$f\xi$	$-\beta u$	$\nabla \cdot (fV)$ (B)	$-kD$ (D)	$-\alpha \nabla^2 P$	$\frac{\partial w}{\partial x}\frac{\partial u}{\partial z} + \frac{\partial w}{\partial y}\frac{\partial v}{\partial z}$
1	5.6	0	0	0	−11.0	0	11.6	11.6	−0.6	0	
2	10.5	0	0	−0.4	−10.1	−0.1	11.8	11.7	−1.6	0	
3	9.6	0	0	−0.5	−9.1	−0.2	11.9	11.7	−2.6	0	
4	8.7	−0.1	0	−0.7	−8.1	−0.4	12.0	11.6	−3.5	0	
5	7.8	−0.1	0	−0.5	−7.2	−0.6	12.0	11.4	−4.2	0	+
6	6.9	−0.2	0.1	−0.8	−6.2	−0.8	12.0	11.2	−5.0	0	
7	6.1	−0.2	0.1	−0.9	−5.3	−1.1	12.0	10.9	−5.6	0	
8	5.3	−0.2	0.2	−0.9	−4.4	−1.4	11.9	10.5	−6.1	0	
9	4.6	−0.2	0.2	−1.1	−3.5	−1.7	11.8	10.1	−6.6	0	
10	3.8	−0.3	0.3	−1.0	−2.8	−2.1	11.8	9.7	−6.9	0	+
11	3.2	−0.3	0.4	−1.1	−2.0	−2.4	11.7	9.3	−7.3	0	
12	2.6	−0.3	0.4	−1.3	−1.2	−2.8	11.5	8.7	−7.5	0	
13	2.1	−0.4	0.6	−1.4	−0.5	−3.2	11.4	8.2	−7.7	0	
14	1.5	−0.4	0.7	−3.2	2.0	−3.6	11.3	7.7	−7.9	−1.8	
15	−26.1	−0.2	0.8	−22.7	4.0	−4.1	11.1	7.0	−5.3	−5.7	−
16	−49.6	0	1.0	−45.3	3.3	−4.6	11.0	6.4	−0.2	−9.5	
17	−45.0	−0.1	1.2	−43.3	−1.5	−5.0	11.0	6.0	4.3	−9.7	
18	−41.5	−0.4	1.2	−41.0	−1.3	−5.3	10.9	5.6	8.5	−14.4	
19	−39.0	−1.0	1.3	−36.3	−2.7	−5.6	10.9	5.3	12.4	−15.0	
20	−37.2	−1.7	1.3	−33.1	−3.7	−5.8	10.9	5.1	16.1	−17.5	−
21	−36.2	−2.5	1.4	−32.9	−2.2	−5.9	10.9	5.0	19.8	−22.6	
22	−35.8	−3.5	1.3	−31.4	−2.2	−6.0	11.0	5.0	23.4	−26.2	
23	−36.0	−4.5	1.3	−31.1	−1.7	−6.1	11.0	4.9	27.0	−30.2	
24	−36.8	−6.0	1.3	−31.8	−0.3	−6.1	11.1	5.0	30.7	−35.4	
25	−38.2	−7.5	1.2	−25.3	−6.6	−6.0	11.1	5.1	34.6	−33.1	−
26	−40.4	−9.5	1.2	−18.6	−13.5	−6.0	11.2	5.2	38.7	−30.4	
27	−43.5	−11.8	1.1	−8.0	−24.8	−5.8	11.3	5.5	43.0	−23.7	
28	−47.2	−14.6	0.8	−1.0	−32.7	−5.5	11.3	5.8	47.8	−20.9	
29	−52.9	−18.0	0.7	0	−50.8	−5.1	11.4	6.3	53.2	−8.7	−

According to Table 6 the largest positive contribution to the substantial derivative of the divergence arose from the Coriolis terms, not surprisingly the βu portion. Since the pressure-force term was zero, the βu term largely determined the entire variation of the divergence along the trajectory leading to the general shape of Curve D in Figure 63. Stated somewhat differently, given Curve A (the acceleration) and B (the Coriolis terms) (Figure 63), Curve D, by (104), follows, that is, the divergence must increase along the trajectory.

By contrast, the next pair of parcels released at 2° and 4°N determined quite a different pattern, largely the result of a non-zero pressure-force term. Indeed up to about 172.5°E the general shape of Curves A, B, and D are quite similar in both Figures 63 and 64. However, beyond about 172°E the pressure-force term in Figure 64 increased rather sharply, and this largely determined the subsequent behavior of Curve D, the divergence. This is easily deduced if we let the acceleration term (Curve A) between 172.5 and 170.5 be zero. Then, appealing to (104) and the independently determined Curves B and C, Curve D, the divergence, must decrease. Beyond 170.5°E the acceleration could no longer be assumed zero, but increased largely in response to the substantial derivative of the horizontal divergence (Table 7) and Equation (100). To the right of Figure 64, that is, further downstream, the divergence soon reached a minimum and began to increase again, thus displaying the spatial lag between the pressure-force term and the divergence already noticed in cases from the Indian Ocean when the divergence of the pressure-force term was also positive.

How is the above analysis altered by including the fifth and sixth terms of Equation (99), the vertical-motion terms? For each pair of parcels we performed five sample computations—roughly for every degree of longitude—using our values of the divergence to compute the vertical motion at 500 m, assuming a vertical wind shear of 1 m $\sec^{-1}$. In the case of the parcels released at the equator and 2°N, the complete vertical-motion term averaged -1.5×10^{-11}; that is, it operated to decrease the divergence of the acceleration, or, more practically in terms of our computing the deformation as a residual, altered the latter by about $\pm 0.4 \times 10^{-11} \sec^{-2}$ on the average. Stated otherwise, the vertical-motion term was about an order of magnitude smaller than the Coriolis terms—the largest single term—and acted in the same sense, that is, our computations of the substantial derivative of the horizontal divergence are very slight underestimates.

In the case of the second pair of parcels released at 2° and 4°N, the vertical-motion terms ranged from 0 to $10 \times 10^{-11} \sec^{-2}$, thus, again making our computations of the substantial derivative of the horizontal divergence slight underestimates; or, again in terms of our residual deformations, slightly increasing the latter, by, say, 15%.

Thus, the significant difference in the horizontal divergence between the above pair of adjacent parcels results from a divergent pressure field; or more familiarly, the curvature of the pressure profile is capable of producing significant differences in the horizontal divergence along adjacent trajectories.

VI. *Some Comments on Errors*

Since the momentum equation was solved as an initial value problem it is vital to ask how changes in the initial conditions are likely to alter the computed speeds and positions as determined by Equations (54) and (56). (It will be assumed that the initial coordinates are known exactly.) Before offering any specific results, it is useful to consider a simple argument involving just one speed component to see what results might be expected. It is clear from Equation (54) that a small change in u_0 leads to a change in u of the form

$$\delta u = e^{-kt}\, \delta u_0, \tag{107}$$

where, for simplicity, cos ft has been set equal to one. Clearly, as time increases the effect of any change, (δu) in the computed speed, resulting from a change in the initial speed (δu_0), rapidly decreases. Specifically if $k = 2 \times 10^{-5}$ sec^{-1} and $u_0 = 3$ m sec^{-1}, then the resulting change in the computed speed (δu) is about one-third of the initial change (δu_0) in something like 10 hours. Thus, if the "error" in the initial speed was 3 m/sec, the corresponding "error" in the computed speed is about 1 m/sec in 10 hours.

Five sample parcels were released along an arbitrary meridian between 5°N and 25°N under a constant-pressure force of 1 millibar per 5 degrees of latitude. Hourly speeds and positions were determined from Equations (54) and (56) for a period of 130 hours, which spatially corresponds to distances of the order of 10 degrees of latitude. For convenience, the parcels were given an initial speed equal to that of the geostrophic wind; subsequently, the initial parcel speeds were varied by 1, 3, and 5 m sec^{-1} above and below the geostrophic value—still under the same constant pressure-gradient force—with the result depicted in Figure 65. The graphs display speed versus time along selected trajectories with the expected result suggested by Equation (107), namely, that the computed speeds are tolerably insensitive to the initial speeds after periods of, say, 10–15 hours; or, after distances of the order of 2 or 3 degrees of latitude. Further, it is clear that the format of the above figures is sensible only if the resulting trajectories traverse substantially the same space. This is essentially the case; the largest spatial difference between the examples, for any time, being about a degree of latitude.

If, on the other hand, the errors in the initial speeds are assigned arbitrarily, say $+3$ m sec^{-1} to one parcel, -2 m sec^{-1} to another, the results are substantially the same as those mentioned above. Because of the close similarity, no diagrams are presented.

The immediate practical consequence of the above results is that the

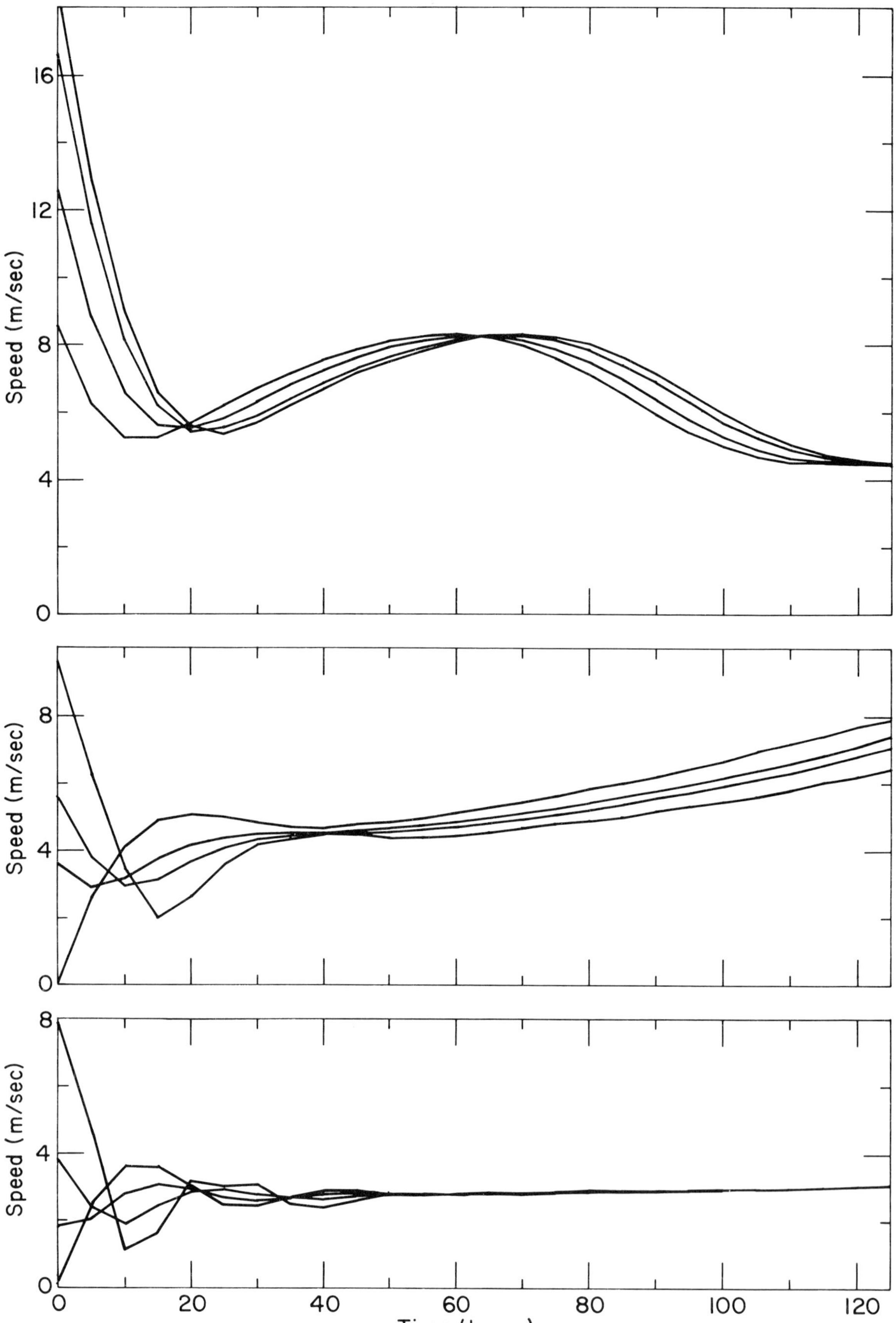

FIGURE 65. Parcel speeds as a function of time along three different trajectories (from top to bottom: near 20°N, 10°N, and 5°N) as a function of different initial speeds.

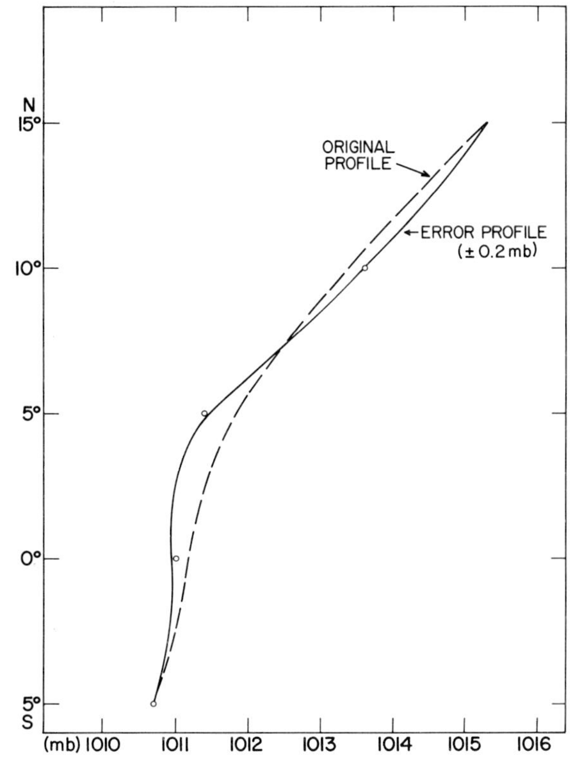

FIGURE 66. Pressure errors assumed in the original January pressure profile for the Indian Ocean (Figure 28). The errors were assumed to be either 0 or ± 0.2 mb at every 5 degrees of latitude between 15°N and 5°S.

initial state is not terribly critical in terms of our diagnostic approach. Nevertheless we cannot afford to be casual; rather, we must decide what we expect to recover from a particular experiment. If we wish to determine just the sign of the divergence developed by a given family of trajectories, then we can allow some margin of error in the initial speeds. If, on the other hand, we require greater detail, then we must know the initial speeds more accurately. For example, almost any initial speeds assigned to the parcels in the examples of the Indian Ocean would have given comparable results at great distances from their initial positions; however, features within 5 degrees of the initial positions would have been altered. Even with initial speeds of zero, parcels released south of the equator in the Indian Ocean in July recover the large convergent area north of the equator.

Appealing to an argument similar to that developed above, it is easy to see that, from Equations (54) and (56), a small change in the gradient P_y leads to changes in the x coordinate and speed given approximately by

$$\delta x = f(k - f)\frac{e^{-kt}}{k^2 + f^2}\delta P_y - \frac{ft}{k^2 + f^2}\delta P_y \tag{108}$$

and

$$\delta u = (1 + e^{-kt})\frac{f}{k^2 + f^2}\delta P_y \tag{109}$$

where, for convenience, both circular functions have been set equal to 1, and $v_0 = 0$. For given values of k and f, the first term in Equation (108) largely determines the change in x for a given change in P_y; at later times the second term dominates. Roughly speaking at 10°N, with $k = 2.0 \times 10^{-5}$ sec^{-1}, and $P_y = 6 \times 10^{-3}$ cm sec^{-2} (which corresponds to a relative error of about 35 percent if $P_y = .017$), the contribution of the second term doubles between 10 and 20 hours, while that of the first term roughly halves during the same period. In particular, x is about 0.5 and 1.0 degree in 10 and 20 hours respectively. The analogous variation in the speed, u, is, from (54), about 2.2 m sec^{-1} and 1.9 m sec^{-1} in 10 and 20 hours respectively. Since a variable Coriolis term is incorporated into the trajectory computations, further estimates of either speed or position based upon Equations (108) and (109) may not be too fruitful; indeed, the more pressing question is how pressure-gradient errors affect the divergence computations.

It is difficult to provide a single experiment that will cover all possibilities—the combinations are endless. Nevertheless, consider the following: adopt the January Indian Ocean pressure profile and assume that an observing station is located every 5 degrees of latitude; let each station have an even chance of having either ± 0.2 mb or zero error; with the new profile (Figure 66) what divergence will ensue from the resulting

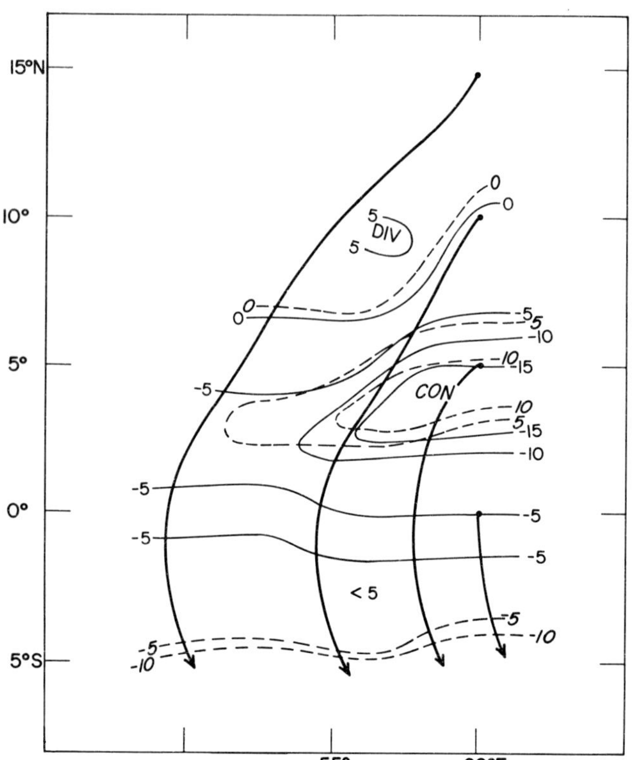

FIGURE 67. Comparison between divergence pattern computed with original January pressure profile and the error profile of Figure 66; values are $\times 10^{-6}$ sec^{-1}.

trajectories? The comparison is presented in Figure 67 and displays little significant change; the spatial variations of the "new" trajectories are barely discernible on the scale of the figure. Obviously, we are not suggesting that pressure errors are not "important"; indeed, small errors in regions of weak gradients could be ruinous. Rather, our collective experience suggests that pressure errors of a few tenths of a millibar (in regions of non-zero gradient) over distances of, say, 5 degrees of latitude coupled with speed errors of, say, a few meters per second are likely to preserve our diagnostic results, that is, we are likely to infer at least the proper sign of the divergence from computed trajectories.

VII. *Concluding Remarks*

Can we now answer the questions raised in the Introduction? Can we understand how air parcels in the surface layer, moving across nearly zonal isobars, are able to undergo such striking changes in their horizontal divergence in comparatively short distances (as in Figure 29, for example)?

The results of the foregoing section indicate that the divergence of both the pressure-gradient force and the Coriolis forces are significant in determining the temporal change of the horizontal velocity divergence "following the motion," for example, (105). Some embroidery is added when the remaining terms of Equation (105) are incorporated, but, in the cases considered here, both pressure gradient and Coriolis forces acting either separately or combined, are never unimportant.

Further, we believe that successful diagnostic results are not possible unless the pressure field is known to within a few tenths of a millibar over distances of 5 degrees of latitude. On the other hand, the initial wind speeds can be known rather poorly (compared to the pressure), a direct consequence of our frictional constraint and the particular way it enters our analytical solutions, in Equations (54) and (56). Without friction these solutions would be extremely sensitive to the initial conditions, as was pointed out above.

We advance no claims of the likely validity of our formulation of friction over continental regions; our results are deemed valid for oceanic regions only.

Some of our computational procedures are clearly capable of refinement, together with at least one change in the physical assumptions. For example, it would be desirable to have a "smoother" description of the pressure field, with the object of eliminating the discontinuities in the pressure gradient employed here. Though, as noted above, the consequences of the latter feature can be effectively suppressed by ignoring a few computed values near such discontinuities, a polynomial representation of the pressure field could eliminate such features *ab initio*.

One additional computational refinement could entail the direct evaluation of the deformation term in (105), rather than deducing it as a residual as we have done here. Then we could evaluate the vertical motion terms in (105) with greater confidence and make the interpretative content of both Equations (38) and (39) a little more exact.

One of the physical assumptions that could bear further scrutiny is that of friction simply opposing the motion. How would our results be altered, for example, if the frictional force acted at some finite angle to the motion and its value were chosen to simulate variations in static stability as reflected in, say, the sea surface temperature as suggested by Bunker's (*92*) results?

Finally, to what heights do the effects of near-surface convergence extend? It is clearly impossible to answer this question solely in the context of our diagnostic results. Indeed, this is currently a subject of active research wherein the CISK (Conditional Instability of the Second Kind) mechanism, amongst others, has been proposed to "parameterize convection" (*93*) in low latitude models of tropical motions. Here we simply note that our diagnostic studies of near-surface convergence provide a necessary prerequisite for a more complete treatment of the convection problem and, in our judgement, with better spatial definition in the divergence field than can normally be computed from the wind field alone.

List of Mathematical Symbols

A	area
a	radius of the earth
B	$\frac{1}{\rho}\frac{\partial \tau_s}{\partial z}$
b	constant
C	$kP_x + fP_y$
D	divergence
Def^2	$\left(\frac{\partial v}{\partial x} + \frac{\partial u}{\partial y}\right)^2 + \left(\frac{\partial u}{\partial x} - \frac{\partial v}{\partial y}\right)^2$
f	Coriolis parameter
g	acceleration of gravity
H	height of top of frictional layer
h	height of aircraft
i	square root of -1
K	the coefficient of eddy viscosity
k	coefficient of frictional resistance
M	$\frac{kP_x + fP_y}{f^2 + k^2}$
m	$fu - P_y$
N	$\frac{kM - P_x}{f}$
P	pressure
P_r	$-\frac{1}{\rho}\frac{\partial P}{\partial r}$
P_ϕ	$-\frac{1}{\rho}\frac{\partial P}{\partial \phi}$
P_x	$-\frac{1}{\rho}\frac{\partial P}{\partial x}$
P_y	$-\frac{1}{\rho}\frac{\partial P}{\partial y}$
p	static stability parameter
q	$u(f^2 + k^2) - C$
R	radius
r	coordinate in polar coordinate system, radius of curvature

s	distance along the flow
t	time
u	component of the wind along the x axis
u_g	component of the geostrophic wind along the x axis
u_0	initial value of the zonal component of the wind
V	tangential velocity
V_{agn}	ageostrophic wind normal to the flow
V_{ags}	ageostrophic wind along the flow
V_n	velocity normal to the flow
V_0	initial scalar wind speed
V_r	radial velocity
V_s	velocity along the flow
v	component of wind along the y axis
v_g	component of the geostrophic wind along the y axis
w	vertical velocity
x,y,z	Cartesian coordinates, positive toward the east, north, and up
α	angle between the wind and the isobars
α^2	proportionality constant for the pressure gradient varying with latitude
λ	coordinate of longitude
μ	coefficient of eddy viscosity
Ω	angular velocity of the earth's rotation
ϕ	coordinate of latitude
ρ	density
ψ	angle of change of direction of flow along axis normal to the flow
θ	angle coordinate in polar coordinate system between x axis and radius vector
τ_x	shearing stress along the x axis
τ_y	shearing stress along the y axis
τ_s	shearing stress along the flow
ξ	vorticity about the local vertical

In addition, the usual "dot" notation for time derivations is used throughout, while the "naught" subscript indicates initial values.

Notes and References

1. Abercromby, Ralph, 1885: *Principles of forecasting by means of weather charts.* 2nd ed., official no. 60. Meteorological Council, Her Majesty's Stationary Office (HMSO), London, 123 pp.
2. Euler, L., 1755: Principles généraux du mouvement des fluides. *Historie de l'Acad. roy. des sciences*, **11**, 274–361.
3. Lagrange, J. L., 1788: *Mécanique analytique.* Paris, 370 pp.
4. Sommerfeld, A., 1950: *Mechanics of Deformable Bodies.* Academic Press, New York, 396 pp.
5. Gerstner, Franz, 1804: Theorie der wellen. *Abhad. Koenigl.*, vol 1., *bohemischen Gesell. Wissenschaften zu Prague*, pp. 176–195.
6. Milne-Thomson, L. M., 1956: *Theoretical Hydrodynamics.* 3rd ed., Macmillan, New York, p. 406.
7. Brandes, H. W., 1824: *Beiträge zur Witterungskunde.* Barth, Leipzig, 411 pp.
8. Espy, James P., 1841: *Philosophy of storms*, Charles C. Little and James Brown, Boston, 552 pp.
9. Shaw, Sir Napier, and R. G. K. Lempfert, 1906: *The life history of surface air currents: a study of the surface trajectories of moving air*, Meteorological Office 174, HMSO, London, 107 pp.
10. Coriolis, G., 1835, 1836: Sur les équations du mouvement relatif des systemes de corps. *J. l'École Polytechn.*, **15**, 142–154; *Comptes* Rend., **2**, 172–174.
11. Poisson, S. D., 1838: Mémoire sur le mouvement des projectiles dans l'air en ayant égard le rotation de la terre. *J. de l'École Polytechn.* **16**, 1–68.
12. Von Baeyer, J. J., 1858: Über die Bahnlinien der Winde auf der spharoidischen Erdoberflache. *Ann. Phys. Chem.*, **104**, 377–404.
13. Ohlert, Bernhard, 1860: Zur theorie der Stromungen des Meeres und der Atmosphäre. *Ann. Phys. Chem.*, **110**, 234–264.
14. Mousson, Albert, 1866: Über das Sieden einer rotireden Flussigkeit. *Ann. Phys. Chem.*, **129**, 168–174.
15. Sprung, A., 1881: Über die Bahnlinien eines freien Theilcheus auf der rotireden Erdoberflache und deren Bedentung für die Meteorologie. *Ann. Phys. Chem.*, **14**, 128–149.
16. Cordeiro, F. J. B., 1910: *The atmosphere: its characteristics, and dynamics.* Spon and Chamberlain, New York, pp. 39–48.
17. Gold, E., 1911: Review of Cordeiro's book (16), *Quart. J. Roy. Meteoro. Soc.*, **37**, 199–202.
18. Whipple, F. J. W., 1917: The motion of a particle on the surface of a smooth rotating globe. *Phil. Mag.*, **33**, 457–471.
19. Shaw, Sir Napier, 1903: On curves representing the paths of air in a special type of traveling storm. *Mon. Wea. Rev.*, **31**, 318–320.
20. Hudson, W. H. H., 1907: Anemoids. *Reports of the British Association* (76th meeting, York, 1906), John Murray, London, pp. 483–485.
21. Gold, E., 1908: *Barometric gradient and wind force.* M. O. 190, HMSO, London, 44 pp.
22. Shaw, Sir Napier, 1913–14: Principia atmospherica—a study of the circulation of the atmosphere. *Proc. Roy. Soc. Edinburgh*, **34**, 77–112.
23. Meteorological Office, 1916: *Meteorological glossary.* HMSO, London, 210 pp.
24. Mohn. H., 1870: *Norwegian atlas of storms.* Quoted in Peslin (*25*).
25. Peslin, H., 1877: On the relation between barometric variations and the general atmospheric currents. *Annual Report Smithsonian Inst.* (1877), Washington, pp. 465–478.
26. Sprung, A., 1885: *Lehrbuch der Meteorologie.* Hoffmann und Campe, Hamburg, 407 pp.

27. Gold, E., 1908: The relation between wind velocity at one thousand meters altitude and the surface pressure distribution. *Proc. Roy. Soc.*, **80A**, 436–443.
28. Sandström, J. W., 1911: On the relation between atmospheric pressure and wind. *Bull. Mt. Weather Obs.*, **3**, 275–303.
29. Gulberg, C. M., and H. Mohn, 1910: Studies on the movements of the atmosphere (1876). *In* C. Abbe, (ed.), *Smithsonian Misc. Coll.*, **51**, 122–248.
30. Exner, F. M., 1913: Über oszillatorische Stromungen der Luft. *Ann. Hydrog. Marit. Meteoro.*, **41**, 145–150.
31. Exner, F. M., 1917: *Dynamische Meteorologie*. Teubner, Leipzig, 308 pp.
32. Exner, F. M., 1925: *Dynamische Meteorologie*. Teubner, Leipzig, 421 pp.
33. Hesselberg, T., 1914: Die Reibung in der Atmosphäre. *Met. Zeit.*, **31**, 220–232.
34. Forsythe, George E., 1949: Exact particle trajectories for nonviscous flow in a plane with a constant Coriolis parameter. *J. Meteoro.*, **6**, 337–346.
35. Meisinger, C. L., 1919: The constant elevation free-balloon flights from Fort Omaha. *Mon. Wea. Rev.*, **47**, 535–538.
36. Richardson, L. F., and D. Proctor, 1925: Diffusion over distances ranging from 3 to 86 kilometers. *Mem. Roy. Meteoro. Soc.*, **1**, 1–16.
37. Ackerman, J. D., and J. F. Piccard, 1937: Upper air study by means of balloons and radio meteorograph. *J. Aero. Sci.* **4**, 332–337.
38. Angell, J. K., 1961: Use of constant-level balloons in meteorology. *Advances in Geophysics*, **8**, 137–219.
39. Bjerknes, V., 1930: Über die hydrodynamischen Gleichungen in Lagraneschen und Eulerscher Form und ihre Linearisierung für das Studium kleiner Storungen. *Geofys. Publik.*, **5**, 43 pp.
40. Baur, F., and H. Philipps, 1938: Untersuchungen der Reibung bei Luftstromungen über dem Meer. *Ann. Hydrog. Marit. Meteoro.*, **66**, 279–302.
41. Philipps, H., 1939: Die Abweichung vom geostrophischen Wind. *Meteoro. Zeits.*, **56**, 460–483.
42. Lotz, Karl, 1940: Über Trajektorien der Luft und deren Divergeny. *Ann. Hydrog. Marit. Meteoro.*, **68**, 227–248.
43. Koschmieder, H., 1933: *Dynamische Meteorologie*, Akademische Verlagsgels., Leipzig, 376 pp.; 2 Auflage (1941), Leipzig, 384 pp.
44. Ekman, V. K., 1905: On the influence of the earth's rotation on ocean currents. *Ark. Math. Astron. Fysik*, **2**, 1–53.
45. Franceschini, Guy A., 1955: *Computation of horizontal trajectories in the atmosphere*; with an Appendix by J. C. Freeman, Jr. Scientific Reports no. 1 and no. 5, A&M Projects 57 and 108, Dept. Oceanography and Meteorology, A&M College of Texas, College Station, 18 pp.
46. Franceschini, Guy A., 1958: *The theory of analysis and prediction of trajectories*. Final Report on Project 108, Dept. of Oceanography and Meteorology, A&M College of Texas, College Station, 94 pp.
47. Kao, S-K, 1960: Nonlinear oscillations of a particle in a long wave in a rotating fluid. *J. Geophy. Res.*, **65**, 279–386.

    1961: An analysis of the trajectory of a constant pressure balloon in a long atmospheric wave in comparison with the non-linear theory. *J. Meteoro*, **18**, 251–256;

    1962: Non-linear oscillations of a particle in a long wave in a rotating viscous fluid. *Phys. Fluids*, **5**, 639–643.
48. Wurtele, M. G., 1960: Lagrangian dynamics and the parcel method in atmospheric models. *J. Meteoro.*, **17**, 661–669.
49. Wiin-Nielsen, A., 1959: On the application of trajectory methods in numerical forecasting. *Tellus*, **11**, 180–196.
50. Djuric, D., 1961: On the accuracy of air trajectory computations. *J. Meteoro.*, **18**, 597–605.
51. Okland, Hans, 1961: An experiment in numerical integration of the barotropic equation by a quasi-Lagrangian method. *Geofys. Publik.*, vol. 22, **5**, 22 pp.
52. Gordon, A. H., and P. M. Shaw, 1954: Application of particle dynamics to derive a general circulation in low latitudes. *Arch. Meteoro. Geophy. Bioklimat.*, **6A**, 319–333.
53. Rayleigh, Lord, 1945: *Theory of Sound* (1877). Vol. 2. Dover, New York, p. 102.
54. Phillips, N. A., 1956: The general circulation of the atmosphere: a numerical experiment. *Quart. J. Roy. Meteoro. Soc.*, **82**, 123–164.
55. Kao, S-K., 1962: Non-linear oscillations of a particle in a long wave in a rotating viscous fluid. *Phys. Fluids*, **5**, 639–643.
56. Gray, William M., 1967: Global view of the origin of tropical disturbances and storms. *Atmospheric Science Paper*, no. 114, Dept. Atmospheric Science, Colorado State University, Fort Collins, pp. 56–62.

57. Swinbank, W. C., 1970: Structure of wind and the shearing stress in the planetary boundary layer. *Arch. Met. Geophys. Biokl.*, **19A,** 1–12.
58. Charnock, H., J. R. D. Francis, and P. A. Sheppard, 1956: An investigation of wind structure in the trades—Anegada, 1953. *Phil. Trans.*, **249A**, 179–234.
59. Roll, H. U., 1965: *Physics of the marine atmosphere.* Academic Press, New York, p. 214.
60. Grimes, A., 1938: The movement of air across the equator. *Memoires of the Malayan Meteoro. Service*, **2**, 11 pp.
61. Treloar, H. M., 1945: An approximate conditional near-surface wind-pressure gradient relation in low latitudes. *Tropical Wea. Res. Bull.*, no. 12, Royal Australian Air Force, Brisbane, pp. 3–27.
Gibbs, W. J., 1945: Motion of air in low latitudes. *Tropical Wea. Res. Bull.*, no. 12, Royal Australian Air Force, Brisbane, pp. 28–37.
Crossley, A. F., 1946: *The motion of air in equatorial latitudes.* M. O. 490c, Memorandum of the Conference of Empire Meteorologists, HMSO, London.

Crossley, A. F., 1948: On the relation between wind and pressure. *Quart. J. Roy. Meteoro. Soc.*, **74**, 379–382.
62. Grimes, A., 1951: Equatorial meteorology. *In* T. F. Malone, ed., *Compendium of Meteorology*, American Meteorological Society, Boston, pp. 881–886.
63. Wyrtki, Klaus, 1956: The computation of oceanic and meteorological fields of motion with friction proportional to the velocity. *Penjelidikan Laut di Indonesia*, **2**, 26 pp.
64. Johnson D. H., and H. T. Mörth, 1959: Value of contour analysis in equatorial meteorology. *Nature*, **184**, 1372–1373.
65. Johnson, D. H., and H. T. Mörth, 1960: Forecasting research in East Africa. *In* D. J. Bargman, ed., *Tropical Meteorology in Africa*, Munitalp Foundation, Nairobi, pp. 56–133.
66. Mörth, H. T., 1964: Five years of pressure analysis over tropical Africa. In J. W. Hutchings, ed., *Proceedings of the Symposium on Tropical Meteorology (Rotorua, November 1963)*, New Zealand Meteorological Service, Wellington, pp. 329–338.
67. Thompson, B. W., 1965: *The Climate of Africa.* Oxford University Press, New York, 8 pp., 132 charts.
68. Ballif, Jae R., 1958: Orders of magnitude of some meteorological variables observed in the Marshall Islands (June 1956). *In* J. R. Ballif, C. E. Palmer, P. C. Sinclair, and W. Viezee, *An empirical study of air movement near the equator*, Final Report under Contract No. AF19(604)–2134, Institute of Geophysics, UCLA, Los Angeles, pp. 10–31.
69. Kao, S-K., and M. Neiburger, 1959: Oscillations and trajectories of air particles in some pressure systems. *J. Geophy. Res.*, **64**, 1283–1291.
70. If Div $\mathbf{V} = 1/A\, dA/dt = 1/y\, dy/dt$, then if, say $v_0 = 0$, and $y = 1/f\,[(u_0 - u_g)(1 - \cos ft)]$, $\dot{y} = -(u_0 - u_g)\sin ft$, (for $k = 0$), it is easy to show that

$$\frac{\partial}{\partial u_0}(\text{Div } \mathbf{V}) = \frac{f^2 \sin ft}{1 - \cos ft} \cdot \frac{1}{u_0 - u_g}\left(\frac{\partial u_0}{\partial y_0}\right)^{-1}.$$

71. Miller, Forrest R., and R. N. Keshavamurthy, 1967: *Structure of an Arabian Sea summer monsoon system.* International Indian Ocean Expedition Meteorological Monographs, no. 1, East-West Center Press, Honolulu, 104 pp.
72. Joint Task Force Seven, 1958: *Meteorological Report on Operation Hardtack* (20 March–31 July 1958). 6 vols., Meteorological Center, Joint Task Force Seven, Pearl Harbor, Hawaii.
73. The mean surface pressure was obtained from the daily observations as published by Joint Task Force Seven (*72*). However, the mean value for Tarawa led to a labored analysis when incorporated with similar mean values from Ocean Island, Arorae, and Funafuti. In fact, the mean pressure reported by the New Zealand Meteorological Service (*Annual Meteorological Summary for Fiji, Tonga and Western Pacific High Commission Territories*, Suva, 1958, p. 20) from their station on the same island (Betio), using Kew-pattern mercurial barometers, was 1.0 mb higher (1009.2) than that deduced from (*72*). Further, these discrepancies persisted for the remaining part of the test period (May–July 1958); the mean values from (*72*) were 1008.0, 1008.3, and 1008.6, respectively, while the New Zealand station reported 1008.7, 1009.0, 1009.4 for the same months. Again, both means were based upon the same four synoptic observations per day (00, 06, 12, 18 GMT) and taken on the same island of the atoll.

Part of the discrepancy could be the result of errors in coding and checking; perhaps as much as 0.3 mb from faulty sea-level reductions (corresponding to height error of 10 feet), but it is difficult to account for a whole mb in the *mean* pressure solely from such sources. Rather, it seems more likely that a major part of the discrepancy is the result of the use of aneroid barometers (whose scales are rarely graduated to less than 0.5 mb) by personnel of the 6th Weather Squadron (Mobile). A recent note (21 June

1968) from the squadron asserts that "one of the observers who participated in the project remembers that aneroid barometers were used." Further, the squadron was unable to provide any specific information regarding calibration procedures employed during the Hardtack operation.

Since the same personnel operated similar stations at both Kusaie and Kapingamarangi, this raises further doubt regarding the quality of the pressure analysis. Though the analysis was extended to include stations south of the equator and found to be consistent, the region is quite open and other interpretations are entirely possible. It, therefore, follows that the computed trajectories in the region of Kusaie and Kapingamarangi must be accepted with some reservations.

74. We attempted to develop a dimensional argument relating k, the frictional coefficient, to the static stability, wind speed, and wind shear in various combinations. Since the undetermined coefficient was nearly as large as some of the other terms, the exercise was not convincing.
75. Flohn, H., 1960: The structure of the intertropical convergence zone: *In* D. J. Bargman, ed., *Tropical Meteorology in Africa*, Munitalp Foundation, Nairobi, pp. 244–251.
76. Sinclair, P. C., 1958: Mean vertical atmospheric motion over the Marshall Islands during June 1956. *In* J. R. Ballif, C. E. Palmer, P. C. Sinclair, and W. Viezee, *An empirical study of air movement near the equator*, Final Report under Contract No. AF19(604)–2134, Institute of Geophysics, UCLA, Los Angeles, pp. 40–56.
77. Koninklijk Nederlands Meteorologisch Instituut, 1958: *General air circulation, wind roses, isobars for the Indian Ocean, Jan., July*. De Bilt.
78. Sadler, James C., 1969: *Average Cloudiness in the Tropics from Satellite Observations*. International Indian Ocean Expedition Meteorological Monographs, no. 2, East-West Center Press, Honolulu, 32 pp.
79. Raman, C. R. V., personal communication.
80. Seelye, C. J., 1950: Rainfall and its variability over the Central and Southwestern Pacific. *New Zealand J. Sci. and Techn.*, Sect. B, **32**, 11–24.
81. New Zealand Meteorological Service, 1964: *Annual meteorological summaries*, Suva, Fiji, 60 pp.
82. This expedient would not be useful if the meridional distribution of initial speeds contained a pronounced extrema; no speed extrema are suggested by the available data, though, admittedly, the latter are rather crude.
83. New Zealand Meteorological Service, 1964: *Daily Weather Bulletin* (July 1964), Wellington, 82 pp.

U. S. Weather Bureau, 1964: *National Summary* (July 1964), Asheville, 75 pp.

84. There is little doubt about the latter feature, no matter how questionable the double trough may be. During 1–8 May 1957 daily sea-level pressure measurements at five stations in the Line Islands (Palmyra, 6°N; Fanning, 4°N; Christmas, 2°N; Malden, 3°S; Penrhyn, 9°S) revealed high pressure (1015 mb) on the equator, after a continuous increase of 10 mb in seven days.
85. Bennett, Edward B., 1963: An Oceanographic Atlas of the Eastern Tropical Pacific, based on data from EASTROPIC expedition, October-December 1955. *Bull. Inter-Amer. Trop. Tuna Comm.*, **8**, 33–165.
86. Neuman, G., and R. E. Williams, 1964: Observations of the equatorial undercurrent in the Atlantic Ocean at 15°W during Equalant I.
87. Ramage, C. S.: Climate of the Indian Ocean. *In* H. Thomsen, ed., *World Survey of Climatology*, vol. 15, Elsevier, New York (in press).
88. Ramage, C. S., F. R. Miller, and Charmian Jefferies, 1972: Meteorological Atlas of the International Indian Ocean Expedition, vol. I: The surface climate of 1963 and 1964, xv pp. + 144 charts, National Science Foundation, Washington. (See chart 31). *J. Geophy. Res.*, **70**, 297–304.
89. Sherman, Leon, 1952: On the scalar vorticity and horizontal divergence equations. *J. Meteoro.*, **9**, 359–366.
90. Gerber, R., 1950: Sur une réduction à un principe variationnel des équations du mouvement d'un fluide visqueux incompressible. *Ann. de l'Institut Fourier*, **1**, 157–162.
91. Riehl, H., 1954: *Tropical Meteorology*. McGraw-Hill, New York, p. 86.
92. Bunker, Andrew F., 1971: Energy transfer and tropical cell structure over the Central Pacific. *J. Atmos. Sci.*, **28**, 1101–1116.
93. Charney, J. G., and A. Eliassen, 1964: On the growth of the hurricane depression. *J. Atmos. Sci.*, **21**, 68–75.